Contents

Chapter 1.	The Band Theory of Solids	6
1.1.	Structure of Atoms	7
1.1.1.	Hydrogen Atom	7
1.1.2.	Bohr's Postulates	7
1.1.3.	Atomic Radii of Orbits and Energy Levels	8
1.1.4.	Quantum Numbers	9
1.1.5.	Quantum Numbers as the Electron Address in an Atom	12
1.2.	Many-Electron Atoms	13
1.2.1.	Pauli's Exclusion Principle	13
1.2.2.	Distribution of Electrons over the Shells	13
1.3.	Degeneracy of Energy Levels in Free Atoms, Removal of Degeneracy by External Effect	16
1.3.1.	Degenerate state	16
1.3.2.	Degeneracy, Removed by an External Field	16
1.4.	Formation of Energy Bands in Crystals	18
1.4.1.	Splitting of Energy Levels in a Crystal	18
1.4.2.	Allowed and Forbidden Bands	19
1.5.	Filling of Energy Bands by Electron	20
1.5.1.	Filled Levels Create Filled Bands While Empty Levels Form Empty Bands	20
1.5.2.	Overlapping of Energy Bands in a Crystal	21
1.6.	Division of Solids into Conductors, Semiconductors, and Dielectrics	22
1.6.1.	Conductors	23
1.6.2.	Semiconductors and Dielectrics	23
1.6.3.	Energy Band Occupancy and Conductivity of Crystals	24
Chapter 2.	Electrical Conductivity of Solids	25
2.1.	Bonding Forces in a Crystal Lattice	25
2.1.1.	Crystal as a System of Atoms in Stable Equilibrium State	25
2.1.2.	Repulsive and Attractive Forces	25
2.1.3.	Covalent bond	27
2.1.4.	Not Every Two Hydrogen Atoms May Form a Molecule	28
2.1.5.	Semiconductors as Typical Covalent Crystals	29

2.2.	Electrical Conductivity of Metals	30
2.3.	Conductivity of Semiconductors	33
2.4.	Intrinsic Semiconductors	33
2.4.1.	Electron Conductivity	34
2.4.2.	Hole Conductivity	36
2.4.3.	The Number of Holes Equals the Number of Free Electrons	37
2.5.	Doped (Impurity) Semiconductors	38
2.5.1.	Donor Impurities	38
2.5.2.	Hole Semiconductors	41
2.6.	Effect of Temperature on the Charge Carrier Concentration in Semiconductors	42
2.6.1.	Effect of Temperature on Conductivity of Intrinsic Semiconductors	43
2.6.2.	Impurity Semiconductors	44
2.6.3.	Degenerate Semiconductors	47
2.7.	Temperature Dependence of Electrical Conductivity of Semiconductors	49
2.7.1.	Scattering by Ionized Impurity Atoms	50
2.7.2.	Scattering by Thermal Vibrations	52
2.7.3.	Temperature Dependence of Conductivity of Semiconductors	53
Chapter 3.	Non equilibrium Processes in Semiconductors	55
3.1.	Generation and Recombination of Non equilibrium Charge Carriers	55
3.1.1.	Free Carrier Generation	55
3.1.2.	Free Carrier Recombination	55
3.1.3.	Equilibrium Carriers	56
3.1.4.	Non equilibrium Carriers	56
3.1.5.	Electra-neutrality Condition	57
3.1.6.	Recombination Rate	58
3.1.7.	The Concept of Trapping Cross Section	59
3.1.8.	Types of Recombination	60
3.1.9.	Radiative Recombination	60
3.1.10.	Non radiative Recombination	61
3.1.11.	Trapping Sites and Trapping Centers	61
3.1.12.	Surface Recombination	62
3.2.	Diffusion Phenomena in Semiconductors	62
3.2.1.	Diffusion Current	62
3.2.2.	The Einstein Relation	64
3.2.3.	Holes Pursue Electrons	64
3.2.4.	Diffusion and Recombination	65

3.3. Photo-conduction and Absorption of Light	66
3.3.1. Photo-induced Processes	66
3.3.2. Laws Governing Photo-conductivity	67
3.3.3. Quantum Efficiency	69
3.3.4. Photo-current Spectral Distribution Curve	70
3.3.5. Intrinsic Photo-conductivity	71
3.3.6. Impurity Photo-conductivity	72
3.3.7. Absorption Spectrum	73
3.3.8. Exciton Absorption	73
3.4. Luminescence	74
3.4.1. Photoluminescence	75
3.4.2. Fluorescence	76
3.4.3. Phosphorescence	77
Chapter 4. Contact Phenomena	80
4.1. Work Function of Metals	80
4.1.1. Double Electric Layer	80
4.1.2. Electrical Image Force	81
4.1.3. Total Work Function	81
4.2. The Fermi Level in Metals and the Fermi-Dirac Distribution Function	83
4.2.1. Fermi Level	83
4.2.2. Fermi-Dirac Distribution Function	85
4.2.3. Effect of Temperature	86
4.2.4. Physical Meaning of the Fermi Level	87
4.3. The Fermi Level in Semiconductors	88
4.3.1. Intrinsic Semiconductor	89
4.3.2. Impurity Semiconductors	89
4.3.3. Effect of Temperature on the Position of the Fermi Level	90
4.3.4. The Fermi Level in Degenerate Semiconductors	92
4.3.5. The Fermi Level and Work Function in Semiconductors	93
4.4. The Contact Potential Difference	95
4.5. Metal-to-Semiconductor Contact	98
4.5.1. State of Equilibrium	98
4.5.2. Barrier Layer	100
4.5.3. Bending of Bands	100
4.5.4. Anti-barrier Layers	101
4.6. Rectifier Properties of the Metal-Semiconductor Junction	101
4.6.1. External Potential Difference Applied in the Forward Direction	103
4.6.2. External Potential Difference Acting in the Cut-Off Direction	105

4.6.3.	Current-Voltage Characteristic of the Metal - Semiconductor Junction	106
4.6.4.	Ohmic Contact	106
4.7.	$p-n$ Junction	107
4.7.1.	Methods of Obtaining $p-n$ Junctions	107
4.7.2.	$p-n$ Junction at Equilibrium	110
4.7.3.	Diffusion Current	111
4.7.4.	Contact potential Difference	112
4.7.5.	Band Structure of the $p-n$ Junction at Equilibrium	112
4.7.6.	$p-n$ Junction as a Barrier layer	115
4.8.	Rectifying Effect of the $p-n$ Junction	117
4.8.1.	Reverse Current	117
4.8.2.	Forward Current	119
4.8.3.	Injection of Carriers	120
4.8.4.	Current-Voltage Characteristic of the $p-n$ Junction	121
4.8.5.	Effect of Temperature on the Rectifier Properties of the $p-n$ Junction	122
4.9.	Breakdown of the $p-n$ Junction	123
4.9.1.	Avalanche Breakdown	124
4.9.2.	Tunnel Breakdown	124
4.9.3.	Thermal Breakdown	125
4.9.4.	Surface Breakdown	126
4.10.	The Electric Capacitance of the $p-n$ Junction	126
Chapter 5.	Semiconductor Devices	128
5.1.	Hall Effect and Hall Pickups	128
5.1.1.	Lorentz Force	128
5.1.2.	Hall Effect in the Extrinsic Semiconductors	129
5.1.3.	Practical Application of the Hall Effect	131
5.1.4.	The Hall Effect in Semiconductors with Mixed Conductivity	132
5.1.5.	Hall Effect in Intrinsic Semiconductors	133
5.1.6.	Hall Pickups-and Their Applications	134
5.1.7.	Hall Magnetometers	134
5.1.8.	Heavy-Current Ammeters	134
5.1.9.	Hall Pickups as Signal Transducers	135
5.1.10.	Hall Microphone	136
5.2.	Semiconductor Diodes	136
5.2.1.	Rectifier Diodes	136
5.2.2.	Stabilizer Diodes (Stabilitrons)	137
5.2.3.	Varicaps	139
5.3.	Tunnel Diodes	140

5.3.1.	Manufacturing of the Tunnel Diodes	140
5.3.2.	The $p-n$ Junction between Degenerate Semiconductors	141
5.3.3.	Tunnel Transitions of Electrons in Equilibrium State	141
5.3.4.	Operation of the Tunnel Diode at Forward Bias Voltage	143
5.3.5.	Operation of the Tunnel Diode at the Reverse Bias	145
5.3.6.	Generation of Continuous Oscillations with the Help of the Tunnel Diode	145
5.3.7.	Backward Diode	147
5.4.	Transistors	149
5.4.1.	$p-n-p$ Transistor	149
5.4.2.	Injection of Holes to the Base	151
5.4.3.	Collector Junction	151
5.4.4.	How Does the Transistor Amplify?	153
5.4.5.	Methods of Connecting a Transistor	153
5.4.6.	Common Emitter Connection	153
5.5.	Semiconductor Injection (Diode) Lasers	155
5.5.1.	Photons Create Photons	155
5.5.2.	Population Inversion	156
5.5.3.	Creation of Population Inversion at the Junction Between Degenerate Semiconductors	158
5.6.	Semiconductor at Present and in Future	160

CHAPTER 1

The Band Theory of Solids

When researchers first come across semiconductors, there was a clear-cut division of all solids into two large groups, viz. conductors (including all metals) and insulators (or dielectrics) and these differed in principle in their properties. These new semiconductor materials could not be included in either of these groups. On the one hand, they conducted electric current, although to a much lesser extent than metallic conductors, and on the other, they did not always conduct. Nevertheless, they did conduct electricity and so were named semiconductors (or half conductors).

Later, it was discovered that semiconductors differ from metals both in the way they conduct and in the way external factors influence their conduction. For example, the effect of temperature on conductivity of metallic conductors and semiconductors is quite opposite. In metals, an increase in temperature causes a gradual decrease in conductivity, while the heating of semiconductors results in a sharp increase in conductivity. The introduction of impurities also has different effects on conductivity of metallic conductors and semiconductors. In metals, as a rule, impurities worsen conductivity, while in semiconductors the introduction of a negligibly small amount of certain impurities can raise the conductivity by tens or even hundreds of thousands of times.

Finally, if we send a beam of light of a flux of some particles onto a conductor, it will have practically no effect on its conductivity. On the other hand, irradiation or bombardment of a semiconductor causes a drastic increase in its conductivity.

It is interesting to note that these properties of semiconductors are, to a considerable extent, typical of dielectrics, hence it would be much more correct to call semiconductors semi-insulators or semi dielectrics.

In order to explain the behavior of semiconductors in various conditions, to account for their properties and to predict new effects, we must consider their structural peculiarities. That is why we shall start with the discussion of the atomic structure of matter.

1.1. Structure of Atoms

1.1.1. Hydrogen Atom. From the course of physics you should know that an atom consists of a nucleus and electrons rotating around it. This model of an atom was proposed by the English physicist Rutherford. In l913, the Danish physicist Niels Bohr, one of the founders of quantum mechanics, used this model for the first correct calculations of hydrogen atom that agreed well with experimental data. His theory of the hydrogen atom has played an extremely important role in the development of quantum mechanics, though it underwent considerable changes later.

1.1.2. Bohr's Postulates. According to the Rutherford-Bohr model, hydrogen atom consists of a singly charged positive nucleus and one electron rotating around it. To a first approximation, it can be assumed that the electron moves along the trajectory which is a circle with the fixed nucleus at its center. According to the laws of classical electrodynamics, any accelerated motion of a charged body (including the electron) must be accompanied by the emission of electromagnetic waves. In the model under consideration, the electron moves with a tremendous centripetal acceleration, and therefore it should continuously emit light. Should it do so, its energy would gradually decrease and the electron would come closer and closer to the nucleus. Finally, the electron would unite with the nucleus ("fall" on it). Nothing of this kind occurs in reality, and atoms do not emit light in their unexcited state. In order to explain this fact, Bohr formulated two postulates.

According to Bohr's first postulate, an electron can only be in an orbit for which its angular momentum (i.e. the product of the electron momentum mv by the radius r of the orbit) is a multiple of $h/2\pi$ (where h is Planck's constant). While the electron is in one of these orbits, it does not emit energy. Each allowed electron orbit corresponds to a certain energy, or certain energy state of the atom, which is called a stationary state. Atoms do not emit light in a stationary state. The analytic expression of Bohr's first postulate is

$$mvr = n\frac{h}{2\pi}$$

where $n = 1, 2, 3........$ is an integer called the principal quantum number.

Bohr's second postulate states that absorption or emission of light by an atom occurs during transitions of the atom from one stationary state to another. The energy is absorbed or emitted upon transition in certain amounts, called quanta, whose value hv is determined by the difference in energies corresponding to the initial and final stationary states of the

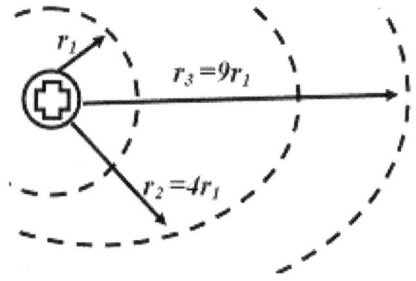

FIGURE 1

atom:
$$h\nu = W_m - W_n$$

where W_m is the energy of the initial state of the atom, W_n the energy of its final state, and ν the frequency of light emitted or absorbed by the atom.

If $W_m > W_n$, the atom emits energy, and if $W_m < W_n$ the energy is absorbed. Quanta of light are called photons.

Thus, according to Bohr's theory, the electron in an atom cannot change its trajectory gradually (continuously) but can only "jump" from one stationary orbit to another. Light is emitted just when the electron goes from a more distant stationary orbit to a nearer stationary orbit.

1.1.3. Atomic Radii of Orbits and Energy Levels. The radii of allowed electron orbits can be found by using Coulomb's law, the relations of classical mechanics, and Bohr's first postulate. They are given by the following expression:

$$r = n^2 \frac{h^2}{4\pi^2 m e^2}$$

The nearest to the nucleus allowed orbit is characterized by $n = 1$. Using the experimentally obtained values of m, e and h, we find for the radius of this orbit

$$r_1 = 0.53 \times 10^{-8} cm$$

This value is taken for the radius of the hydrogen atom. Any other orbit with a quantum number n has the radius

$$r_n = n^2 r_1$$

Hence, the radii of successive electron orbits increase as n^2 (Fig. 1).

The total energy of an atom with an electron in the nth orbit is given by the formula

$$W_n = -\frac{2\pi^2 m e^4}{n^2 h^2}$$

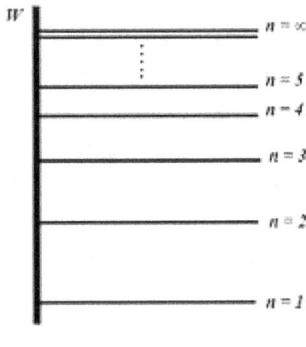

FIGURE 2

These energy values are called atomic energy levels. If we plot the possible values of energy of an atom along the vertical axis, we shall obtain the energy spectrum of the allowed states of an atom (Fig. 2)

It can be seen that with increasing n, the separation between successive energy levels rapidly decreases. This can be easily explained: an increase in the energy of an atom (due to the energy absorbed by the atom from outside) is accompanied by a transition of the electron to more remote orbits where the interaction between the nucleus and the electron becomes weaker. For this reason, a transition between neighboring far orbits is associated with a very small change in energy. The energy levels corresponding to remote orbits are so close that the spectrum becomes practically continuous. In the upper part the continuous spectrum is bounded by the ionization level of the atom ($n = \infty$) which corresponds to the complete separation of the electron from the nucleus (the electron becomes free).

The minus sign in the expression for the total energy of an atom indicates that atomic energy is the lower the closer is the electron to the nucleus. In order to remove the electron from the nucleus, we must expend a certain amount of energy, i.e. supply a definite amount of energy to the atom from outside. For $n = \infty$, i.e. when the atom is ionized, the energy of an atom is taken equal to zero. This is why negative values of energy correspond to $n \neq \infty$. The level with $n = 1$ is characterized by the minimum energy of the atom and the minimum radius of the allowed electron orbit. This level is called the ground, or unexcited level. Levels with $n = 2, 3, 4......$ are called excitation levels.

1.1.4. Quantum Numbers. According to Bohr's theory electrons move in circular orbits. This theory provided good results only for the simplest atom, viz. the hydrogen atom. But it could not provide quantitatively correct results even for the helium atom. The next step was the planetary model of an atom. It was assumed that electrons, like the planets of the solar system, move in elliptical orbits with the nucleus at one of the foci. However, this model was also soon exhausted since it failed to answer many questions.

This is connected with the fact that it is impossible in principle to determine the nature of the motion of an electron in an atom. There are no analogues of this motion in the macro-world accessible for observation. We are not only unable to trace the motion of an electron but we cannot even determine exactly its location at a particular instant of time. The very concept of an orbit or the trajectory of the motion of an electron in an atom has no physical meaning. It is impossible to establish any regularity in the appearance of an electron at different points of space. The electron is "smeared" in a certain region usually called the electron cloud. For an unexcited atom, for example, this cloud has a spherical shape, but its density is not uniform. The probability of detecting the electron is highest near the spherical surface of radius r_1, corresponding to the radius of the first Bohr orbit. Henceforth, we shall assume that the electron orbit is a locus of points which are characterized by the highest probability of detecting the electron or, in other words, the region of space with the highest electron cloud density.

The electron cloud will be spherical only for the unexcited state of the hydrogen atom for which the principal quantum number is $n = 1$ (Fig. 3 a) . When $n = 2$, the electron, in addition to a spherical cloud whose size is now four times greater, may also form a dumb-bell-shaped cloud (Fig. 3 b). The nonsphericity of the region of predominant electron localization (electron cloud) is taken into account by introducing a second quantum number l, called the orbital quantum number. Each value of the principal quantum number n has corresponding positive integral values of the quantum number l from zero to $(n-1)$:

$$l = 0, 1, 2, \ldots\ldots(n-1)$$

For example, when $n = 1$, l has a single value equal to zero. If $n = 3$, l may assume the values $0, 1$ and 2. For $n = 1$ the only orbit is spherical, therefore $l = 0$. When $n = 2$, both the spherical and the dumb-bell-shaped orbits are possible, hence l may be equal either to zero or unity. For $n = 3$, $l = 0, 1, 2$. The electron cloud corresponding to the value $l = 2$ has quite a complicated shape. However, we are not interested in the shape of the electron cloud but in the energy of the atom corresponding to it.

The energy of the hydrogen atom is only determined by the value of the principal quantum number n and does not depend on the value of the orbital number l. In other words, if $n = 3$, the atom will have the

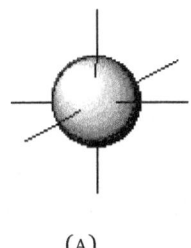 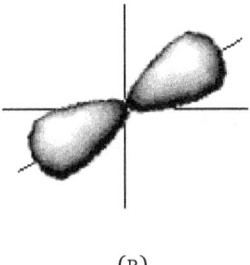

(A) (B)

FIGURE 3

energy W_3, regardless of the shape of the electron orbit corresponding to the given value of n and various possible values of l. This means that upon a transition from the excitation level to the ground level, the atom will emit photons whose energies are independent of the value of l.

While considering the spatial model of an atom, we must bear in mind that electron clouds have definite orientations in it. The position of an electron cloud in space relative to a certain selected direction is defined by the magnetic quantum number m, which may assume integral values from $-l$ to $+l$, including 0. For a given shape (a given value of l), the electron cloud may have several different spatial orientations. For $l = 1$, there will be three, corresponding to the $-1, 0$ and $+1$ values of the magnetic quantum number m. When $l = 2$, there will be five different orientations of the electron cloud corresponding to $m = -2, -1, 0, 1$ and $+2$. Since the shape of the electron cloud in a free hydrogen atom does not influence the energy of the atom, the more so it applies to the spatial orientation.

Finally, a more detailed analysis of experimental results revealed that electrons in the orbits may themselves be in two different states determined by the direction of the electron spin. But what is the electron spin?

In 1925, English physicists G. Uhlenbeck and S. Goudsmit put forward a hypothesis to explain the fine structure of the optical spectra of some elements. They suggested that each electron rotates about its axis like a top or a spin. In this rotation the electron acquires an angular momentum called the spin. Since the rotation can be either clockwise or anticlockwise, the spin (in other words, the angular momentum vector) may have two directions. In $h/2\pi$ units, the spin is equal to $1/2$ and has either " $+$ " or " $-$ " sign depending on the direction. Thus, the electron orientation in the orbit is determined by the spin quantum number σ equal to $\pm 1/2$. It should be noted that the spin orientation like the orientation of the electron orbit, does not affect the energy of hydrogen atom in a free state.

Subsequent investigations and calculations have shown that it is impossible to explain the electron spin simply by its rotation about the axis. When the angular velocity of the electron was calculated, it was found that the linear, velocity of points on the electron equator (if we assume that the electron has the spherical shape) would be higher than the velocity of light, which is impossible. The spin is an inseparable characteristic of the electron like its mass or charge.

1.1.5. Quantum Numbers as the Electron Address in an Atom. Thus, we have learned that ,in order to describe the motion of the electron in an atom or, as physicists say, to define the state of an electron in an atom, we must define, a set of four quantum numbers: n, l, m, and σ.

Roughly speaking, the principal quantum, number n defines the size of the electron orbit, The larger n, the greater region of space is embraced by the corresponding electron cloud. By setting the value of n, we define the number of the electron shell of the atom. The number n, itself can acquire any integral value from 1 to ∞

$$n = 1, 2, 3, ...$$

The orbital quantum number l define the shape of the electron cloud. From the entire set of orbits corresponding to the same value of n., the orbital number l selects those having the same shape. To each value of l there corresponds its own sub shell. The number of sub shells is equal to n, since l may acquire the values from 0 to $(n-1)$:

$$l = 0, 1, 2.....(n-1)$$

The magnetic quantum number m defines the spatial orientation of the orbit in group of orbits with the same shape, i.e. belonging to the same sub shell. In each sub shell, there are $(2l+1)$ orbits with different orientations, since m may assume the values from 0 to $\pm l$:

$$m = -l, -(l-1), ... -1, 0, +1, ..., +(l-1), +l$$

Finally, the spin quantum number σ defines the orientation of the electron spin in the given orbit. Spin has only two values:

$$\sigma = \pm 1/2$$

While considering the hydrogen atom and using the concepts of "shell", "sub shell" and "orbit", we spoke about the opportunities available to the single electron in this atom rather than about the atomic structure. The electron in the hydrogen atom may go from one shell to another and from orbit to orbit within the same shell.

The pattern of electron distribution in many electron atoms and their possible transitions are much more complicated.

1.2. Many-Electron Atoms

1.2.1. Pauli's Exclusion Principle. In discussing the structure of many-electron atoms, we must consider a very important principle formulated in 1925 by the Swiss physicist W. Pauli. This principle states that there cannot be two electrons in an atom in the same quantum state described by the set of four quantum numbers (n, l, m, and σ). In other words, only one or two electrons may be simultaneously in any stationary orbit in an atom. In the latter case, the spins of the electrons must have opposite directions, i.e. for one electron $\sigma = +1/2$, while for the other $\sigma = -1/2$.

Taking into account the Pauli exclusion principle and knowing the number of stationary orbits characterized by different quantum numbers, we can determine the possible number of electrons in every atomic shell and sub shell (see Table 1)

1.2.2. Distribution of Electrons over the Shells. The first shell, whose principal quantum number is $n = 1$, does not split into sub-shells, since it has only one quantum number l associated with it and this is equal to zero. In this case $m = 0$ as well, so we may conclude that the first shell consists of only one orbit which can be occupied, according to Pauli's exclusion principle, by only two electrons.

The second shell ($n = 2$) consists of two sub shells since l can be either 0 or 1. In atomic physics, letter symbols instead of the numerical values of l are used for describing sub shells. For example, regardless of the value of the principal. quantum number n, all sub shells with $l = 0$ are denoted by s, sub shells with $l = 1$ are denoted by p, for $l = 2$ the symbol d is used, and so on. In this connection, it is said that the second shell consists of the $s-$ and $p-$sub shells. The $s-$sub shell ($l = 0$) consists of one circular orbit and may contain only two electrons, while the $p-$sub shell consists of three orbits (m may be equal to $-1, 0$, and $+1$) and may contain six electrons. The total number of electrons in the second shell is equal to eight.

Similarly, we can calculate the possible number of electrons in any shell and sub shell. For example, there can be 10 electrons in the $3d-$sub shell ($n = 3, 1 = 2$), viz. two electrons in each of the five orbits characterized by different values of the quantum number m. The maximum number of electrons in any sub shell is equal to $2(2l+1)$. In spectroscopy, letter symbols (terms) are ascribed to different shells: the first shell is denoted by K, the second by L, the third by M, and so on.

Quantum Numbers				Notation for a sub shell level	No of orbits in a sub shell	No of electrons in a level	Total no of electrons in a shell
n	m	l	σ				
1	0	0	$\pm\frac{1}{2}$	$1s$	1	2	2
2	0	0	$\pm\frac{1}{2}$	$2s$	1	2	6
	1	$-1, 0, +1$	$\pm\frac{1}{2}$	$2p$	3	6	
3	0	0	$\pm\frac{1}{2}$	$3s$	1	2	18
	1	$-1, 0, +1$	$\pm\frac{1}{2}$	$3p$	3	6	
	2	$-2, -1, 0, +1, +2$	$\pm\frac{1}{2}$	$3d$	5	10	

TABLE 1

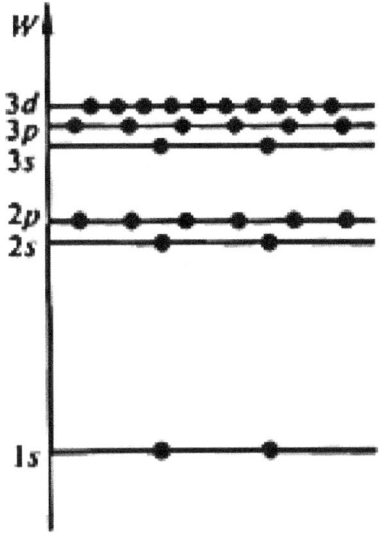

FIGURE 4

The single electron in a hydrogen atom is in a centrally symmetric field of the atomic nucleus; its energy is determined solely by the value of the principal quantum number n and does not depend on the values of the other quantum numbers. On the other hand, in many-electron atoms each electron is in the field created both by the nucleus and by the other electrons.

Consequently, the energy of an electron in many-electron atoms turns out to depend both on the principal quantum number n and on the orbital number l, though remaining independent of the values of m and σ.

This feature of many-electron atoms leads to considerable differences between their energy spectrum and the spectrum of hydrogen atom. Fig.4 shows a part of the spectrum for many-electron atom (the energy levels of the first three atomic shells). Dark circles on the levels indicate the maximum number of electrons which can occupy the corresponding sub shell.

It is well known that a system not subjected to external effect tends to go into the state with the lowest energy. Atom is not an exception in this respect. As the atomic shells are filled, the electrons tend to occupy the lowest levels and would all occupy the first level if there were no limitations imposed by Pauli's exclusion principle. The only electron in the hydrogen atom occupies the lowest orbit belonging to the 1s- level. In the helium atom, the same orbit contains also the second electron, and the first atomic shell is filled. It should be noted that helium is an inert gas, and its great stability is due to the complete outer shell.

In the lithium atom, there are only three electrons. Two of them occupy the first shell, and the third is in the second shell with $n = 2$ (it cannot occupy the first shell due to Pauli's exclusion principle). Lithium is an alkali metal whose valency equals unity. This means that the electron in the second shell is weakly bound to the atomic core and can be easily detached from it. This can be judged from the ionization potential which for lithium is only equal to $5.37\,V$, while for helium it is equal to $24.45\,V$.

As the number of electrons in an atom increases, the outer sub shells and shells are filled. For example, starting with boron, which has 5 electrons, the $2p$-sub shell is filled. This process is completed in inert gas neon which has a fully filled second shell and is thus characterized by the great stability. The eleventh electron in the sodium atom starts populating the third shell ($3s$—sub shell), and so on.

1.3. Degeneracy of Energy Levels in Free Atoms, Removal of Degeneracy by External Effect

1.3.1. Degenerate state. .We have already noted that in many-electron atoms the energy of electrons is only determined by the values of the quantum numbers n and l and does not depend on the values of m and σ. This can be illustrated by the energy spectrum shown in Fig. 4. Indeed, all six electrons in the $3p$-sub-shell, for example, have the same energy, although they have different values of m and σ. States described by different sets of quantum numbers but having the same energy are called degenerate. Similarly, the energy levels corresponding to these states are also called degenerate. The levels are degenerate while the atoms are in the free state. If however, the atoms are placed in a strong magnetic or electric field, the degeneracy is partially or completely removed. Let us illustrate this removal of degeneracy with respect to the quantum number m.

1.3.2. Degeneracy, Removed by an External Field. Different values of the quantum number m correspond to different spatial orientations of similar electron orbits. In the absence of an external field, different orientations of the orbits do not affect the energy of the electrons. If however, we place an atom in an external field, the field will act differently on the electrons in orbits oriented in different ways with respect to the direction of this field. As a result, changes in energies of electrons in similarly shaped but differently oriented orbits will be different both in magnitude and in sign: energies of some. electrons will increase while those of others will decrease. The energy levels for different electrons in the spectrum will also change their arrangement. Moreover, instead of one energy level corresponding to all electrons in similar orbits several

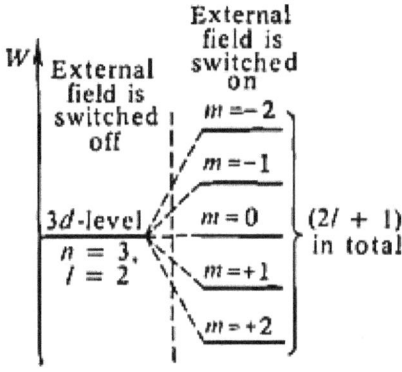

FIGURE 5

sub levels appear in the spectrum, the number of sub level being equal to the number of differently oriented similar orbits, i.e. to the number of possible values of the quantum number m. Fig. 5 shows the result of an external electric field acting on the $3d$—level, for which $n = 3$ and $l = 2$. It can be seen that splitting of the level into sub levels and the displacement of sub levels occur simultaneously.

The process in which previously indistinguishable (from the point of view of energy) degenerate levels become distinguishable is called the removal of degeneracy. Let us illustrate degeneracy removal with another example.

We consider an electron having a certain energy W_0 in a one-dimensional space characterized by the coordinate x (Fig. 6). In the absence of an external field, the state of this electron is described by one energy level W_0 irrespective of the direction of its motion. In other words, in the absence of an external field the energy level W_0 is doubly degenerate. If we apply an external electric field, say, along the x-axis, the energy of the electron becomes dependent on the direction of its motion. If the electron moves along the x-axis, it will be decelerated by the external field, its energy becoming $W_0 - eEx$ (where x is the distance covered by the electron).

If the electron moves in the opposite direction, its energy becomes $W_0 + eEx$. Correspondingly, the appearance of two different states is manifested in the energy spectrum by the slitting of the degenerate level W_0 into two non degenerate levels $W_0 - eEx$ and $W_0 + eEx$. In other words, the degeneracy is removed under the effect of the external field,

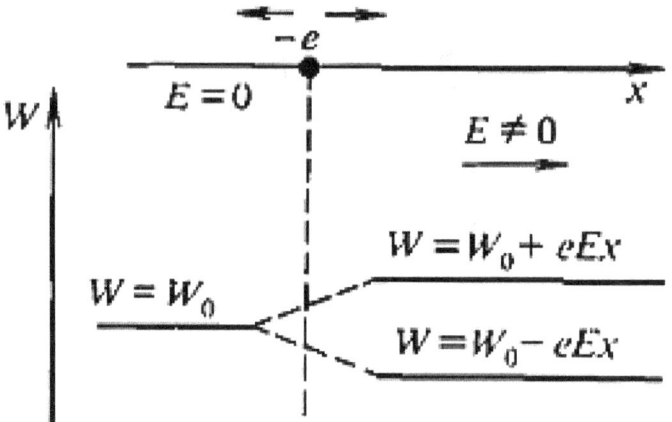

FIGURE 6

1.4. Formation of Energy Bands in Crystals

1.4.1. Splitting of Energy Levels in a Crystal. Let us do the following mental experiment. Take N atoms of a substance and arrange them at a sufficiently large distance from each other but in such a way that this arrangement reproduces the crystalline structure of the material. Since the separation between the atoms is large, we can ignore their interaction and consider them free. In each of these atoms, there are degenerate levels with degeneracies equal to the number of differently oriented similar orbits in corresponding sub shells. Let us now start bringing the atoms closer, retaining their mutual arrangement. As the atoms converge, come closer, they begin to experience the influence of their approaching neighbors, which is similar to the influence of an external electric field. The smaller the separation between the atoms, the stronger is the interaction between them. Owing to this interaction, degeneracy of the energy levels characterizing the free atoms is removed: each degenerate level splits into $(2l + 1)$ non degenerate levels. All the atoms in a crystal generally exist under the same conditions (except for those which form the external boundary of the crystal). It could seem therefore that each atom should contribute the same set of non degenerate sub levels into the energy spectrum that characterized the crystal as a whole viz. one $1s$- sub level, three $2p$-sub levels, five $3d$-sub levels, and so on. Each sub level may contain two electrons with opposite spins. Although this splitting actually occurs, the corresponding sub levels obtained from similar atomic levels differ from each other in energy, some of them are higher in the energy spectrum of the crystal than the

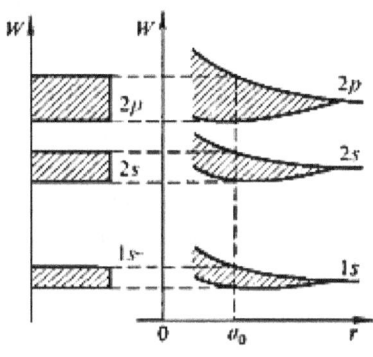

FIGURE 7

initial levels of the individual atoms, while others lie somewhat lower. This difference can be explained by Pauli's exclusion principle generalized for the entire crystal as a single entity. According to this principle, no two non degenerate sub levels in a crystal may have the same energy. Therefore, when the crystal is formed, each energy level spreads into an energy band consisting of $N(2l + 1)$ non degenerate sub levels differing in energy. For example, the $1s$-level spreads into $1s$-band consisting of N sub levels which may contain $2N$ electrons, the $2p$-level spreads into $2p$-band consisting of $3N$ sub levels which may contain $6N$ electrons, and so on.

The formation of energy band in a crystal from discrete energy levels of individual atoms is shown schematically in Fig. 7. The shorter the distance r, the stronger the effect of the neighboring atoms and the more the levels are "smeared". The energy spectrum of a crystal is determined by the smearing of the levels corresponding to the inter atomic distance a_0, typical of a given crystal.

The degree of smearing of levels depends on their depth in an atom. The inner electrons are strongly coupled to their nuclei and are screened from external effects by the outer electron shells. Therefore the corresponding energy levels are weakly smeared. Naturally, the electrons in the outer shells are most strongly affected by the field of the crystal lattice, and the energy levels corresponding to them are smeared the most. It should be noted that smearing of levels into energy bands does not depend on whether there are electrons on these levels or whether they are empty. In the latter case, the smearing of levels indicates the broadening of the range of possible energies which the electron may acquire in the crystal.

1.4.2. Allowed and Forbidden Bands. From what has been said above, it follows that there is an entire band of allowed energy values corresponding to each allowed energy level in a crystal, i.e. there is an allowed band. Allowed bands alternate with the bands of forbidden energy, or forbidden bands. Electrons in a pure crystal cannot have an energy lying in the forbidden bands. The higher the allowed atomic level on the energy scale, the more the corresponding band is smeared. As the energy increases, the forbidden bands becomes narrower.

The separation of sub levels in an allowed band is very small. In real crystals ranging from 1 to 100 cm^3 in size, the sub levels are separated by 10^{-22} to 10^{-24} eV. This difference in energy is so small that the bands are considered to be continuous. Nevertheless, the fact that sub levels in the bands are discrete and the number of sub levels in the band is always finite plays a decisive role in crystal physics, since depending on the filling of the bands by electrons, all solids can be divided into conductors, semiconductors, and dielectrics.

1.5. Filling of Energy Bands by Electron

1.5.1. Filled Levels Create Filled Bands While Empty Levels Form Empty Bands. Since the energy bands in solids are formed from the levels of individual atoms, it is quite obvious that their filling by electrons will be determined above all by the occupancy of the corresponding atomic levels by electrons.

Let us consider by way of an example the lithium crystal. In the free state, the lithium atom has three electrons. Two of these are in the $1s$-shell, which is thus completed. The third electron belongs to the $2s$—sub shell, which is half-filled. Consequently, when a crystal is formed, the $1s$-band turns out to be filled completely, the $2s$- band is half-filled, while the $2p$-, $3s$-, $3p$-, etc. bands in an unexcited lithium crystal are empty, since the levels from which they are formed are unoccupied.

The same is true for all alkali metals. For example, when a sodium crystal is formed, the $1s-$, $2s-$, and $2p-$ bands are completely filled, since the corresponding levels in sodium atoms are completely packed by electrons (two electrons in the $1s$—level, two electrons in the $2s-$ level, and six electrons in the $2p-$ level). The eleventh electron in the sodium atom only half-fills the $3s-$ level, hence the $3s-$ band too is half-filled with electrons.

When crystals are formed by atoms with completely filled levels, the created bands in general are also filled completely. For example, if we constructed a crystal from neon atoms, the $1s-$, $2s-$, and $2p-$ bands in the energy spectrum of such a crystal would be completely filled (each neon atom has 10 electrons which fill the corresponding energy levels).

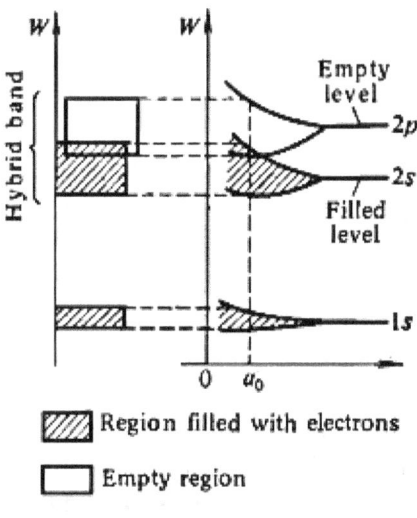

FIGURE 8

The remaining upper-lying bands (3s, 3p, etc.) would turn out to be empty.

1.5.2. Overlapping of Energy Bands in a Crystal. In some cases the problem of filling the energy bands by electrons is more complicated. This refers to crystals of rare-earth elements and those with a diamond-type lattice, among which the most interesting for us are the crystals of typical semiconductors, viz. germanium and silicon.

At a first glance, the crystals of rare-earth elements must only have completely filled and empty bands in their energy spectrum. Indeed, the beryllium atoms, for example, which have four electrons each, are characterized by two completely filled levels, 1s and 2s levels. In magnesium atom, which has 12 electrons, the levels $1s, 2s, 2p$ and $3s$ are also completed. However, the upper energy bands in crystals of the rare-earth elements, which are created by completely filled atomic levels, are in fact only partially filled. This can be explained by the fact that the energy bands corresponding to the upper levels are smeared so much in the process of crystal formation that the bands overlap. As a result of this overlapping, hybrid bands are formed, which incorporate both filled and empty levels. For example, a hybrid band in a beryllium crystal is formed by the completed 2s-levels and the empty 2p-levels (Fig. 8), while in the magnesium crystal, by the filled 3s-levels and empty 3p-levels. It is due to the overlapping that the upper energy bands in rare-earth crystals are filled only partially.

In semiconductor crystals with diamond-type lattices band overlapping leads to quite the opposite result. In silicon atoms, for example, the 3p-level (3p-sub-shell) contains only two electrons, though this level may be occupied by six electrons. It is natural to expect that during the formation of a silicon crystal, the upper energy band (the 3p-band) will only be filled partially, while the preceding band (the 3s-band) will be filled completely (since it is formed by the completely filled 3s-level). Actually the overlapping during the formation of a silicon crystal not only leads to the appearance of a hybrid bands composed of the 3s- and 3p-sub-levels, but also to a further splitting of the hybrid band into two sub bands separated by the forbidden energy gap W_g (Fig. 9). In all, the $3s + 3p$ hybrid band must have 8 electron vacancies per atom (2 vacancies in the 3s-sub-shell and 6 in the 3p-sub-shell). After the splitting of the hybrid band, 4 vacancies per atom are found to be in each sub band. Trying to occupy the lower energy levels, electrons of the third shells of silicon atoms (there are four of them-two in the 3s-sub-shell and two in the 3p-sub-shell) just fill the lower sub band, leaving the upper sub band empty.

1.6. Division of Solids into Conductors, Semiconductors, and Dielectrics

Physical properties of solids, and first of all their electric properties, are determined by the degree of filling of the energy bands rather than by the process of their formation. From this point of view all crystalline bodies can be divided into two quite different groups.

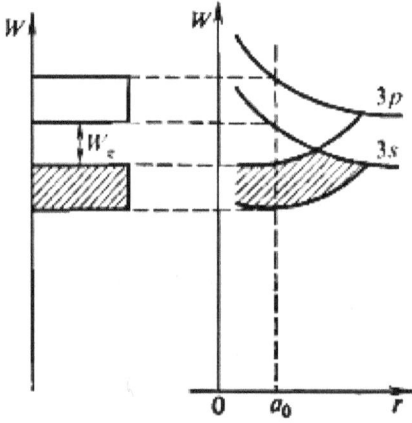

FIGURE 9

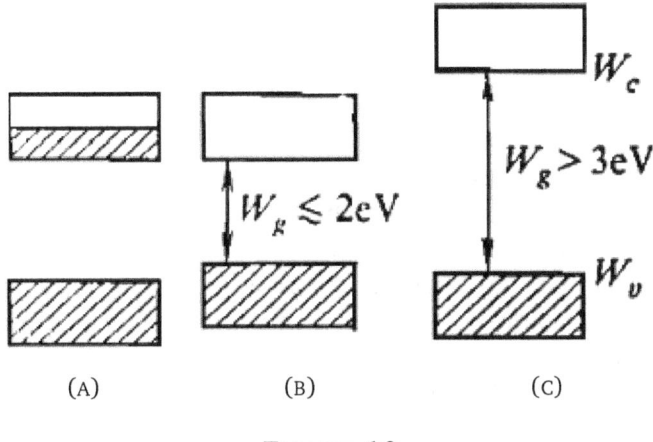

FIGURE 10

1.6.1. Conductors. The first group includes substances having a partially-filled band in their energy spectrum above the completely filled energy bands (Fig. 10 a). As was mentioned above, a partially filled band is observed in alkali metals whose upper band is formed by unfilled atomic levels, and in alkali-earth crystals with a hybrid upper band formed as a result of the overlapping of filled and empty bands. All substances belonging to the first group are conductors.

1.6.2. Semiconductors and Dielectrics. The second group comprises substances with absolutely empty bands above completely filled bands (Fig. 10 b,c). This group also includes crystals with diamond-type structures, such as silicon, germanium, gray tin, and diamond itself. Many chemical compounds also belong to this group, for example, metal oxides, carbides, metal nitrides, corundum (Al_2So_3) and others. The second group of solids includes semiconductors and dielectrics.

The uppermost filled band in this group of crystals is called the valence band and the first empty band above it, the conduction band. The upper level of the valence band is called the top of the valence band and denoted by W_v. The lowest level of the conduction band is called the bottom of the conduction band and denoted by W_c.

In principle, there is no difference between semiconductors and dielectrics. The division in the second group into semiconductors and dielectrics is quite arbitrary and is determined by the width W_g of the forbidden energy gap separating the completely filled band from the empty band. Substances with forbidden band widths $W_g \lesssim 2eV$ belong to the semiconductor subgroup.

Germanium ($W_g \simeq 0.7eV$), silicon ($W_g \simeq 1.2eV$) gallium arsenide $GaAs$ ($W_g \simeq 1.5eV$), and indium antimonide $InSb$ ($W_g \simeq 0.2eV$) are typical semiconductors.

Substances for which $Wg > 3eV$ belong to dielectrics. Well-known dielectrics include corundum ($W_g \simeq 7eV$), diamond ($W_g > 5eV$), boron nitride ($W_g \simeq 4.5eV$), and others.

The arbitrary nature of the division of second group solids into dielectrics and semiconductors is illustrated by the fact that many generally known dielectrics are now used as semiconductors. For example, silicon carbide with its forbidden band width of about $3eV$ is now used in semiconductor devices. Even such a classical dielectric as diamond is being investigated for a possible application in semiconductor technology.

1.6.3. Energy Band Occupancy and Conductivity of Crystals. Let us consider the properties of a crystal with the partially filled upper band at absolute zero ($T = 0$). Under these conditions and in the absence of an external electric field, all the electrons will occupy the lowest energy levels in the band, with two electrons in a level, in accordance with Pauli's exclusion principle.

Let us now place the crystal in an external electric field with intensity E. The field acts on each electron with a force $F = -eE$ and accelerates it. As a result, the electron's energy increases, and it will be able to go to higher energy levels. These transitions are quite possible, since there are many free energy levels in the partially filled band. The separation between energy levels is very small, therefore even extremely weak electric fields can cause electron transitions to upper-lying levels. Consequently, an external field in solids with a partially filled band accelerates the electrons in the direction of the field, which means that an electric current appears. Such solids are called conductors.

Unlike conductors, substances with only completely filled or empty bands cannot conduct electric current at absolute zero. In such solids, an external field cannot create a directional motion of electrons. An additional energy acquired by an electron due to the field would mean its transition to a higher energy level. However, all the levels in the valence band are filled. On the other hand, there are many vacancies in the empty conduction band but there are no electrons. Common electric fields cannot impart sufficient energy for electron to transfer from the valence band to the conduction band (here we do not consider fields which cause dielectric breakdown). For all these reasons an external field at absolute zero cannot induce an electric current even in semiconductors. Thus, at this temperature a semiconductor does not differ at all from a dielectric with respect to electrical conductivity.

CHAPTER 2

Electrical Conductivity of Solids

2.1. Bonding Forces in a Crystal Lattice

2.1.1. Crystal as a System of Atoms in Stable Equilibrium State.
How is a strictly ordered crystal lattice formed from individual atoms? Why cannot atoms approach one another indefinitely in the process of formation of a crystal? What determines a crystal's strength?

In order to answer these questions, we must assume that there are forces of attraction F_{at} and repulsive forces F_{rep} which act between atoms and which attain equilibrium during the formation of a crystal structure. Irrespective of the nature of these forces, their dependence on the interatomic distance turns out to be the same (Fig. 11 a). At the distance $r > a_0$ attractive forces prevail; while for $r < a_0$, repulsive forces dominate. At a certain distance $r = a_0$ which is quite definite for a given crystal, the attractive and repulsive forces balance each other, and the resultant force F_r (which is depicted by curve 3) becomes zero. In this case, the energy of interaction between particles attains the minimum value W_{0r} (Fig. 11 b). Since the interaction energy is at its minimum at $r = a_0$ atoms remain in this position (in the absence of external excitation), because removal from each other, as well as any further approach, leads to an increase in the energy of interaction. This means that at $r = a_0$, the system of atoms under consideration is in stable equilibrium. This is the state which corresponds to the formation of a solid with a strictly definite structure, viz. a crystal.

2.1.2. Repulsive and Attractive Forces. Curve 2 in (Fig. 11 a) shows that repulsive forces rapidly increase with decreasing distance r between the atoms. Large amounts of energy are required in order to overcome these forces. For example, when the distance between a proton and a hydrogen atom is decreased from $r = 2a$ to $r = a/2$ (where a is the radius of the first Bohr orbit), the energy of repulsion increases 300 times. For light atoms whose nuclei are weakly screened by electron shells, the repulsion is primarily caused by the interaction between nuclei. On the other hand, when many-electron atoms get closer, the repulsion is explained by the interaction of the inner, filled electron shells.

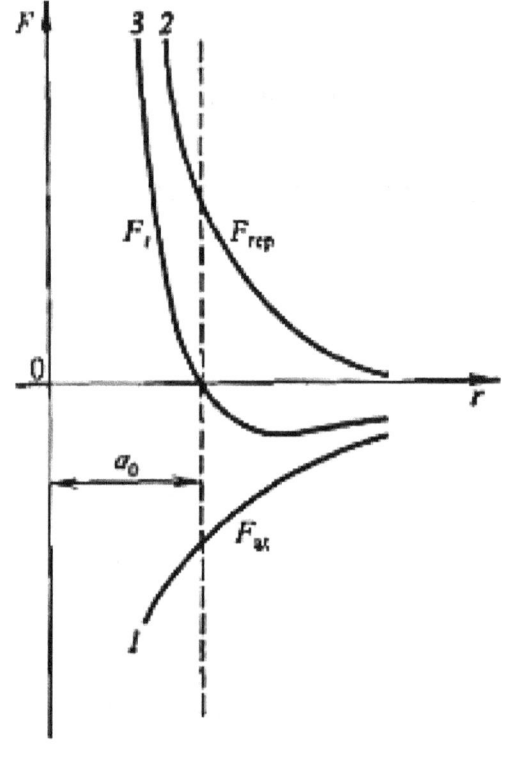

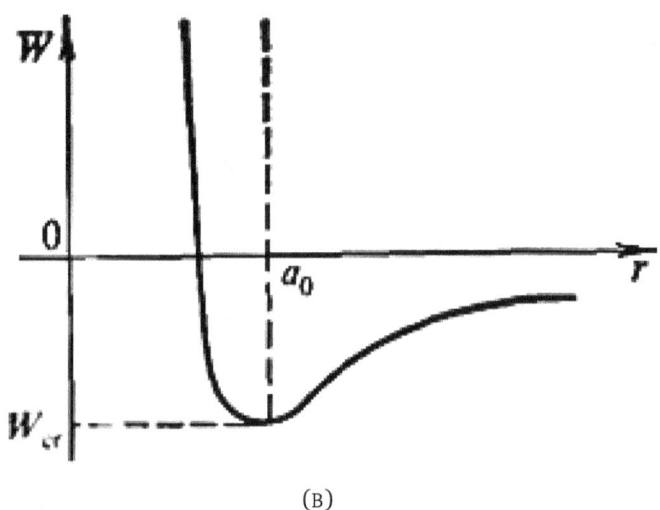

FIGURE 1

The repulsion in this case is not only due to the similar charge of the electron shells but also due to rearrangement of the electron shells. At very small distances, the electron shells should overlap, and orbits common to two atoms will appear. However, since the inner, filled orbits have no vacancies, and extra electrons cannot appear in them due to the Pauli exclusion principle, some of these electrons must go to higher shells. Such a transition is associated with an increase in the total energy of the system, which explains the appearance of repulsive forces.

Obviously, the nature of repulsive forces is the same for all atoms and does not depend on the structure of outer, unfilled shells. On the contrary, forces of attraction which act between atoms are much more diverse in nature, which is determined by the structure and degree of filling of the outer electron shells. Bonding forces acting between atoms are determined by the nature of attractive forces. When considering the structure of crystals, the most important bonds are the ionic, covalent, and metallic, and these should be well known to you from the course of chemistry. Here, we shall only consider the covalent bond, which determines the basic properties of semiconductor crystals.

2.1.3. Covalent bond. Covalent bond is the main one in the formation of molecules or crystals from identical or similar atoms. Naturally, during the interaction of identical atoms, neither electron transfer from one atom to another nor the formation of ions takes place. The redistribution of electrons, however, is very important in this case as well. The process is completed not by the transfer of an electron from one atom to another but by the collectivization of some electrons: these electrons simultaneously belong to several atoms.

Let us see how the covalent bond is formed in the molecule of hydrogen, H_2. Whilst the two hydrogen atoms are far apart, each of them "possesses" its own electron, and the probability of detecting "foreign" electrons within the limits of a given atom is negligibly small. For example, when the distance between the atoms is $r = 5nm$, an electron may appear in the neighboring atom once in $10^{12} years$. As the atoms come closer, the probability of "foreign" electron appearing sharply increases. For $r = 0.2nm$, the transition frequency reaches $10^{12} sec^{-1}$, and at a further approaching the frequency of electron exchange becomes so high that "own" and "foreign" electrons appear at the same frequency near the two nuclei. In this case, the electrons are said to be collectivized. The bond based on the "joint possession" of two electrons by two atoms is called the covalent bond.

The formation of the covalent bond must naturally be advantageous from the point of view of energy. Curve 1 in Fig. 12 shows the total energy of two hydrogen atoms as a function of their separation (the total energy of two infinitely remote hydrogen atoms not interacting with one another is taken as the zero level). It can be seen that as atoms approach one another, their total energy decreases and attains its minimum value at $r = a_0$ which corresponds to the separation of atoms in a hydrogen molecule.

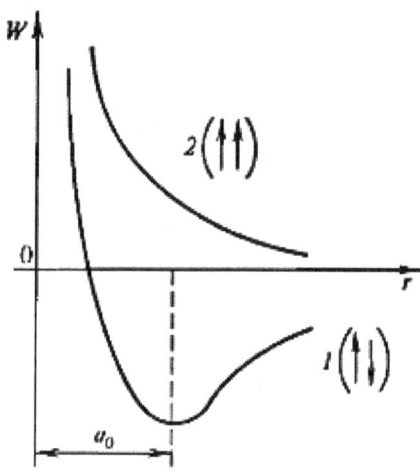

FIGURE 2

2.1.4. Not Every Two Hydrogen Atoms May Form a Molecule. The above process of the formation of the hydrogen molecule is possible only when the electrons in atoms being combined have opposite spins. Only in this case may the electrons occupy the same electron orbit, which is a common orbit for the combined atoms. If, however, the electrons of the approaching hydrogen atoms have parallel spins, i.e. are in the same state determined by the same set of four quantum numbers $(n, 1, m,$ and $\sigma)$, then according to Pauli's exclusion principle they cannot occupy the same orbit. Such atoms will be repelled rather than attracted to one another, and their total energy will increase and not decrease with decreasing r (curve 2 in Fig. 12). Obviously, a molecule is not formed in this case.

An important feature of covalent bonds is their saturation. This property is also one of the manifestations of Pauli's exclusion principle and consists in the impossibility of a third electron to take part in the formation of the covalent bond. Having combined into the covalent bond, electrons are unable to form another bond and at the same time they "do not allow" other electrons to penetrate the combined orbit. For this reason hydrogen molecule, for example, consists of two atoms, and the H_3 molecule cannot be formed.

2.1.5. Semiconductors as Typical Covalent Crystals. The diamond-type crystals exhibit the most typical features of covalent bonds. Typical representatives of this group are semiconductor crystals and among them the well known germanium and silicon. Atoms of these elements have four electrons in the outer shell, each of which forms a covalent bond with four nearest neighbors (Fig. 13). In this process, each atom gives its neighbor one of its valence electrons for "partial possession" and simultaneously gets an electron from the neighbour on the same basis.

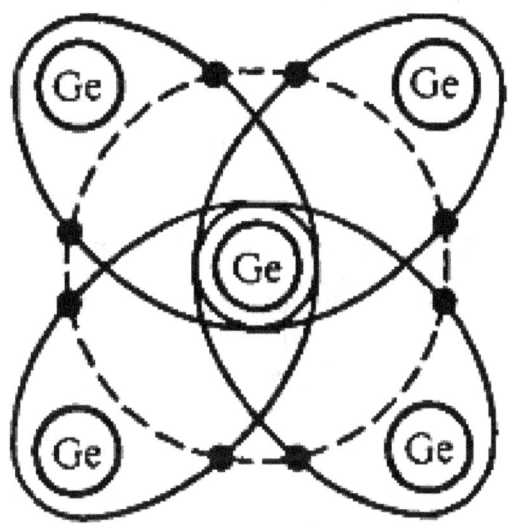

FIGURE 3

As a result, every atom forming the crystal "fills up" its outer shell to complete the population (8 electrons), thus forming a stable structure, which is similar to that of the inert gases (in Fig. 14, these 8 electrons are conventionally placed on the circular orbit shown by the dashed curve). Since the electrons are indistinguishable, and the atoms can exchange electrons, all the valence electrons belong to all the atoms of the crystal to the same extent. A semiconductor crystal thus can be treated as a single giant molecule with the atoms joined together by covalent bonds. Conventionally, these crystals are depicted by a plane structure (Fig. 14). where each double line between atoms shows a covalent bond formed by two electrons.

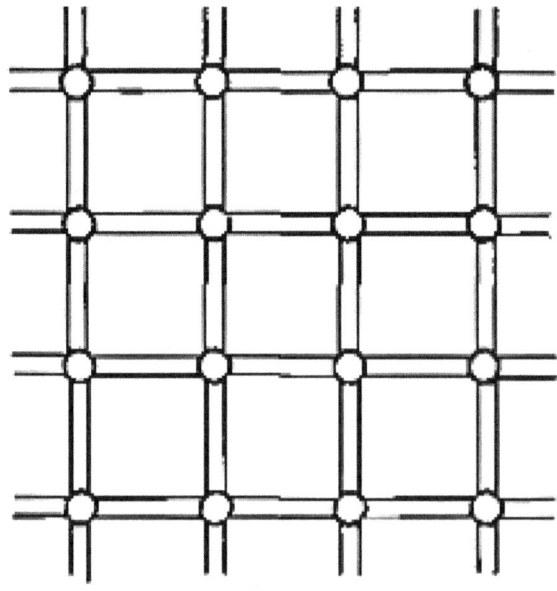

FIGURE 4

2.2. Electrical Conductivity of Metals

The best account of this phenomenon is given by the quantum theory of solids. But to elucidate the general aspects, we can limit ourselves to a consideration based on the classical electron theory. According to this theory, electrons in a crystal can, to a certain approximation, be identified with an ideal gas by assuming that the motion of electrons obeys the laws of classical mechanics. The interaction between electrons is thus completely ignored, while the interaction between electrons and ions of the crystal lattice is reduced to ordinary elastic collisions.

Metals contain a tremendous number of free electrons moving in the interstitial space of a crystal. There are about 10^{23} atoms in 1 cm^3 of a crystal. Hence, if the valence of a metal is Z, the concentration (number density) n of free electrons (also called conduction electrons) is equal to $Z \times 10^{23} cm^{-3}$. They are all in random thermal motion and travel through the crystal at a very high velocity whose mean value amounts to 10^8 cm/sec. Due to the random nature of this thermal motion, the number of electrons moving in any direction is on the average always equal to the number of electrons moving in the opposite direction, hence in the absence of an external field the electric charge carried by electrons is zero. Under the action of an external field, each electron acquires an additional velocity and so all the free electrons in the metal move in the direction opposite to the direction of the applied field intensity. The directional motion of electrons means that an electric current appears in the conductor.

In an electric field of intensity E, each electron experiences a force $F = eE$. Under the action of this force, the electron acquires the acceleration

$$a = \frac{F}{m} = \frac{eE}{m}$$

where e is the charge of an electron and m is its mass.

According to the laws of classical mechanics, the velocity of electrons in free space would increase indefinitely. The same would be observed during their motion in a strictly periodic field (for example, in an ideal crystal with the atoms fixed at the lattice sites).

Actually, however, the directional motion of electrons in a crystal is quite insignificant due to imperfections in the lattice's potential field. These imperfections are mostly associated with thermal vibrations of the atoms (in the case of metals, atomic cores) at the lattice sites, the vibrational amplitude being the larger the higher the temperature of the crystal. Moreover, there are always various defects in crystals caused by impurity atoms, vacancies at the lattice sites, interstitial atoms, and dislocations. Crystal block boundaries, cracks, cavities and other macro defects also affect the electric current.

In these conditions, electrons are continuously colliding and lose the energy acquired in the electric field. Therefore, the electron velocity increases under the effect of the external field only on a segment between two collisions. The mean length of this segment is called the mean free path of the electron and is denoted by λ.

Thus, being accelerated over the mean free path, the electron acquires the additional velocity of directional motion

$$\Delta v = a\tau$$

where τ. is the mean free time, or the mean time between two successive collisions of the electron with defects. If we know the mean free path λ, the mean free time can be calculated by the formula

$$\tau = \frac{\lambda}{v_0 + \Delta v}$$

where v_0 is the velocity of random thermal motion of the electron. Usually, the mean free path λ of the electron is very short and does not exceed 10^{-5} cm. Consequently, the mean free time τ. and the increment of velocity Δv are also small. Since $\Delta v \ll v_0$, we have

$$\tau \simeq \frac{\lambda}{v_0}$$

Assuming that upon collision with a defect the electron loses practically the velocity of directional motion, we can express the mean velocity, called the drift velocity, as follows:

$$\vec{v} = \frac{\Delta v}{2} = \frac{eE}{2m} \cdot \tau = \frac{e\lambda}{2mv_0} E = uE$$

The proportionality factor

$$u = \frac{e}{2m} \frac{\lambda}{v_0}$$

between the drift velocity $\vec{v}$ and the field intensity E is called the electron mobility.

The name of this quantity reflects its physical meaning: the mobility is the drift velocity acquired by electrons in an electric field of unit intensity. A more rigorous calculation taking into account the fact that in random thermal motion electrons have different velocities rather than the constant velocity v_0 gives a double value for the electron mobility:

$$u = \frac{e}{m} \frac{\lambda}{v_0}$$

Accordingly, a more correct expression for the drift velocity is given by the formula

$$\vec{v} = \frac{e\lambda}{mv_0} E$$

Let us now find the expression for the current density in metals. Since electrons acquire an additional drift velocity $\vec{v}$ under the action of an external electric field, all the electrons that are a t a distance not exceeding $\bar{v}$ from a certain area element normal to the direction of the field intensity will pass through i t in a unit of time. If the area of this element is S, all the electrons contained in the parelleopide of length $\bar{v}$; will pass through it in a unit of time (Fig. 15). If the concentration of free electrons in the metal is n, the number of electrons in this volume will be $n\bar{v}S$ The current density, which is determined by the charge carried by these electrons through unit area, can be expressed as follows:

$$j = \frac{en\bar{v}S}{S} = \frac{ne^2\lambda}{mv_0} E$$

The ratio of the current density to the intensity of the field inducing the current is called electrical conductivity and is denoted by σ. Obviously, we get

$$\sigma = \frac{j}{E} = \frac{ne^2\lambda}{mv_0} = neu$$

The reciprocal of electrical conductivity is called resistivity, ρ:

$$\rho = \frac{1}{\sigma}$$

Note that the appearance of an electric current in a conductor is clearly connected, with the electron drift. The drift velocity turns out to be quite low and in real electric fields it usually does not exceed the velocity of a pedestrian. At the same time, current propagates through wires almost instantaneously and can be detected in every part of a closed circuit practically at the same time. This can be explained by the extremely high velocity of propagation of the electric field itself. When a voltage source is connected to a circuit, the electric field reaches the more remote sections of the circuit at the velocity of light and causes the drift of all the electrons at once.

2.3. Conductivity of Semiconductors

As in the case of metals, electric current in semiconductors is related to the drift of charge carriers. In metals the presence of free electrons in a crystal is due to the nature of the metallic bond itself, while in semiconductors the appearance of charge carriers depends on many factors among which the purity of a semiconductor and its temperature are the most important.

Semiconductors are classified as intrinsic, and impurity (extrinsic), or doped. Impurity semiconductors, in turn, can be divided into electronic, or $n-type$ semiconductors and hole, or $p-type$ semiconductors depending on the type of impurity introduced into it. Let us consider each of these groups separately.

2.4. Intrinsic Semiconductors

Intrinsic semiconductors are those that are very pure. The properties of the whole crystal are thus determined only by the properties of atoms of the semiconductor material itself.

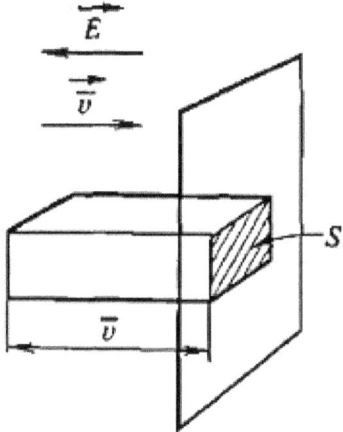

FIGURE 5

2.4.1. Electron Conductivity. At temperatures close to absolute zero, all the atoms of a crystal are connected by covalent bonds which involve all the valence electrons. Although, as we mentioned above, all valence electrons belong equally to all the atoms of the crystal and may go from one atom to another, the crystal does not conduct. Every electron transition from one atom to another is accompanied by a reverse transition. These two transitions occur simultaneously, and the application of an external field cannot create any directional motion of charges. On the other hand, there are no free electrons at such low temperatures.

From the point of view of band theory, this situation corresponds to the completely filled valence band and an empty conduction band.

As the temperature increases, the thermal vibrations of the crystal lattice impart an additional energy to electrons. Under certain conditions, the energy of an electrons becomes higher than the energy of the covalent bond, and the electron ruptures this bond and travels to the crystal interstice, thus becoming "free". Such an electron can freely move in the interstitial space of the crystal independent of the movements of other electrons (electron 1 in Fig. 16).

On the energy levels diagram, the "liberation" of an electron means the electron transition from the valence band to the conduction band (Fig. 17). The energy of rupture of the covalent bond in a crystal is exactly equal to the forbidden band width W_g, i.e. the energy required for an electron to change from a valence electron to a conduction electron. It is clear that the narrower the forbidden gap for a crystal, the lower the temperature at which free electrons begin to appear. In other words, at the same temperature, crystals with a narrower forbidden band will

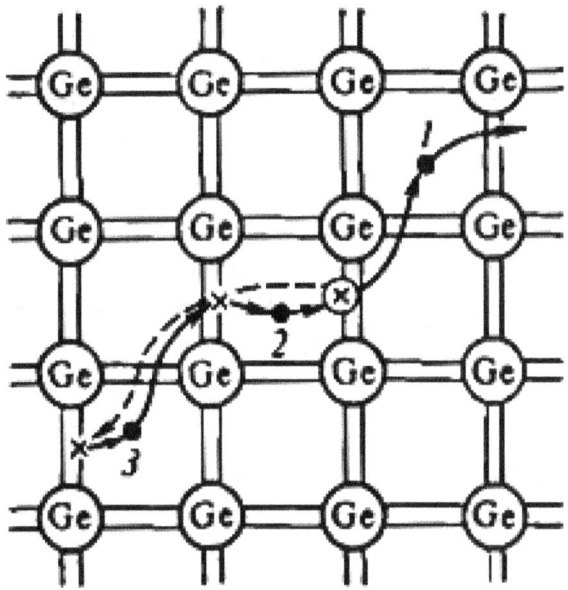

FIGURE 6

have higher conductivity due to a higher electron number density n in the conduction band. Table 2 presents the data on W_g and n for some materials at room temperature.

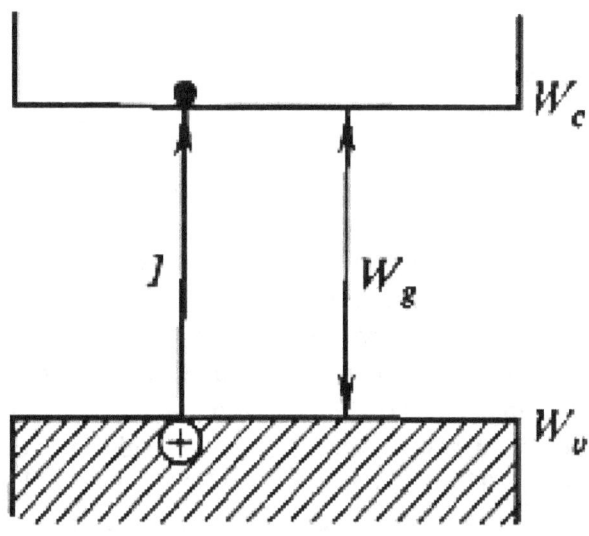

FIGURE 7

Material	$W_g(ev)$	$n\,cm^{-3}$
Indium antimonide	0.2	10^{16}
Germanium	0.7	10^{13}
Diamond	5.0	10^{2}

TABLE 1

If, for example, we heat diamond to 600 K, the number density of free electrons in it will increase so much that becomes comparable with that of the conduction electrons in germanium at room temperature. This is another reason why the division of solids into dielectrics and semiconductors is arbitrary.

2.4.2. Hole Conductivity. A great number of free electrons appearing with increasing temperature is only one of the causes of intrinsic conductivity of a semiconductor.Another cause is associated with a change in the structure of the valence bonds in the crystal, and this is due to a transfer of valence electrons to the interstitial space. Each electron which moves into interstices and becomes a conduction electron leaves a vacancy, or "hole", in the system of valence bonds of the crystal (in Fig. 16 the hole is shown as a light circle; the cross indicates the rupture of the bond caused by this transition). This vacancy may be occupied by a valence electron from any neighboring atom. The vacancy formed as a result of this process may in turn be occupied by an electron from a neighbouring atom, and so on. Such electron transitions to vacant places do not require reverse transitions (as it was in the case of a completely filled system of valence bonds in a crystal), and the possibility of a directional charge transfer appears in the crystal. In the absence of an external field, these transitions are equally possible in all directions, hence the total charge carried through any area element in the crystal is zero. However, when the external field is switched on, these transitions become directional: electrons in the system of valence bonds move in the same direction as free conduction electrons. The movement of electrons in such a transition chain occurs consecutively, as if each electron in turn moves into the vacancy left by its predecessor. If we analyze the result of this consecutive process, it can be treated as the movement of the vacancy itself in the opposite direction.

For the sake of illustration, let us consider a chain of checkers with a vacancy (Fig. 18 a). The consecutive motion of four checkers from left to right (Fig. 18 b) can be considered as the motion of the vacancy itself by four steps in the opposite direction. Something of this kind takes place in a semiconductor. The consecutive transition of electrons 2 and 3 (Fig.

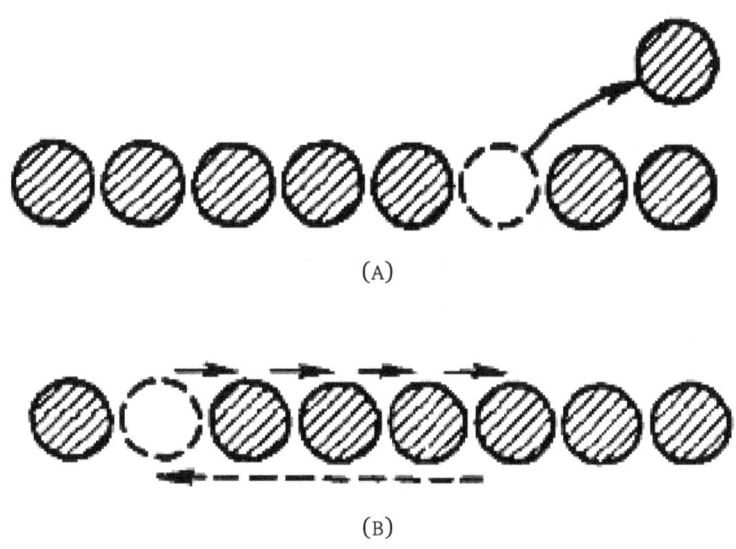

FIGURE 8

16) into the vacancy left by electron 1 is equivalent to the transition of the vacancy in the opposite direction, as shown by the dashed line.

In semiconductor physics, these vacancies are called holes. Each hole is ascribed a positive charge $+e$, which is equal numerically to the electron charge. This approach allows us to consider a series of transitions of a single hole instead of describing the consecutive transitions of a chain of electrons (each to the neighboring atom), and this considerably simplifies our calculations.

The hole conductivity in an intrinsic semiconductor can be explained by the band theory. A transfer of electrons to the conduction band (see Fig. 17) is accompanied by the formation of vacancies (holes) in the valence band, which previously was completely filled. Therefore, electrons remaining in the valence band now can move to vacant higher energy levels. This means that in an external electric field they may acquire an acceleration and thus take part in the directional charge transfer, viz. in creating electric current.

2.4.3. The Number of Holes Equals the Number of Free Electrons. In an intrinsic semiconductor, there are two basic types of charge carriers: electrons (which carry negative charge) and holes (carrying positive charge). The number of holes is always equal to the number of electrons because the appearance of an electron in the conduction band always leads to a hole appearing in the valence band. Hence, electrons and holes are equally responsible for the conductivity of an intrinsic semiconductor. The only difference is that electron conductivity is due to the motion of free electrons in the interstitial space of the crystal (i.e. the motion of electrons traveling to the conduction band), while hole conductivity is associated with a transfer of electrons from atom to atom in the system of covalent bonds of the crystal (i.e. transitions of electrons remaining in the valence bands).

Since there are two types of charge carriers in intrinsic semiconductors, the expression for its electrical conductivity can thus be represented as the sum of two terms:

$$\sigma = en_i u_n + ep_i u_p$$

where n_i is the electron number density in an intrinsic semiconductor, p_i the hole concentration, and u_n and u_p the mobilities of electrons and holes, respectively.

In spite of the apparent equivalence of electrons and holes and the equality of their concentrations, the contribution of electron conductivity to the conductivity of an intrinsic semiconductor is much larger than that of hole conductivity. This is because of the higher mobility of electrons in comparison with holes. For example, the electron mobility in germanium is almost twice the mobility of holes, while in indium antimonide $InSb$ the ratio between the electron and hole mobilities is as much as 80.

Although we shall cover the topic later, note that conductivity in a semiconductor may be caused not only by an increase in temperature but also by other external effects such as irradiation by light or bombardment by fast electrons. What is necessary, is that an external effect causes a transition of electrons from the valence band to the conduction band or, in other words, there must be conditions for generating free charge carriers in the bulk of the semiconductor.

Intrinsic conductivity with the strict equality of number densities of unlike charge carriers ($n_i = p_i$) can only be realized in super-pure ideal semiconductor crystals. In reality we always have to deal with crystals contaminated to some extent. Moreover, it is doped semiconductors that are most important for semiconductor technology.

2.5. Doped (Impurity) Semiconductors

2.5.1. Donor Impurities. The presence of impurity atoms in the bulk of an intrinsic semiconductor leads to certain changes in the energy spectrum of the crystal. While the valence electrons in an intrinsic semiconductor may only have energy in the allowed band region (within

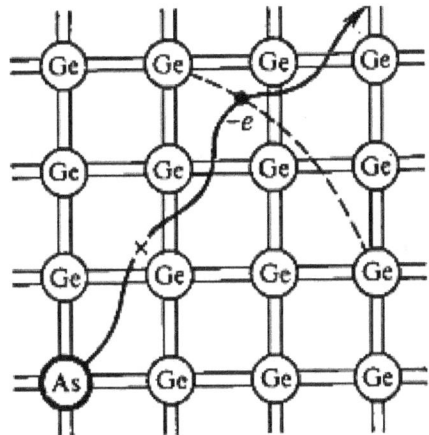

FIGURE 9

the valence band or the conduction band) and their presence in the forbidden band is ruled out, the electrons of some impurity atoms may have energies lying within the forbidden band. Thus, additional allowed impurity levels appear in the energy spectrum in the forbidden band between the top W_v of the valence band and the bottom W_c of the conduction band.

Let us first consider how impurity levels appear by using an electronic (n–type) semiconductor as an example. This is obtained when pentavalent arsenic atoms are introduced as an impurity into a tetravalent germanium crystal (Fig. 19). Four of the five valence electrons of arsenic take part in the formation of covalent bonds with the four nearest neighboring germanium atoms, thus participating in the construction of a crystal lattice. These electrons are in the same conditions as the electrons of the atoms of the parent material (germanium), and thus have the same energy values as the electrons of germanium atoms and lie within the valence band of the energy spectrum. Consequently, these electrons of arsenic atoms do not change the energy spectrum of germanium. The fifth electron, however, does not participate in the formation of covalent bonds. Since it still belongs to the arsenic atom, it continues to move in the field of the atomic core. The interaction between the electron and the atomic core is, however, considerably weakened like the Coulomb force of interaction between two charges placed in a dielectric. The dielectric constant for germanium is $\epsilon = 16$, hence the force of interaction between the arsenic atomic radical and the fifth valence electron is weakened 16 times and the energy of their bond becomes almost 250 times less. Owing to this, the orbital radius of the fifth electron increases 16 times,

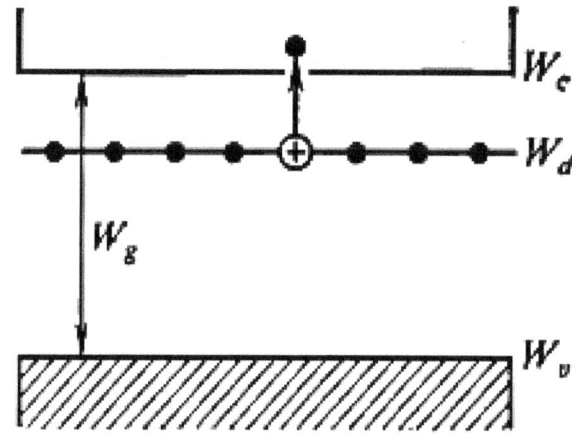

FIGURE 10

and in order to detach it from the atom and make it into a conduction electron, only 0.01 eV energy is required.

In terms of the band theory, this situation just means that an additional allowed level corresponding the energy of the fifth valence electron of the arsenic atom has appeared in the energy spectrum of the crystal. This level lies near the bottom of the conduction band (Fig. 20) and is separated from it by $W_d \simeq 0.01\ eV$.

At temperatures close to absolute zero, all the fifth electrons of the arsenic atoms remain bonded to their atomic cores, in other words, are on their donor levels. The conduction band is therefore empty, and at $T = 0$ an electronic semiconductors does not differ from a typical dielectric as was the case for an intrinsic semiconductor. However, given a slight increase in temperature so that the energy of thermal vibrations of the lattice becomes comparable with the bond energy $W_d \simeq 0.01\ eV$, the fifth electrons are detached from their arsenic atoms and go to the conduction band. The electronic semiconductor acquires conductivity due to free electrons appearing in the interstitial space of the crystal.

It should be emphasized that positive charges that remain after electrons have left the donor levels differ in principle from the holes of intrinsic semiconductors. The escaping electrons of impurity atoms did not take part in the formation of covalent bonds in the crystal nor did they belong to the valence band. Therefore, the remaining positive charges are positively charged ions of the donor impurity (arsenic in the case under consideration), fixed in the crystal lattice and making no contribution to the conductivity of the crystal.

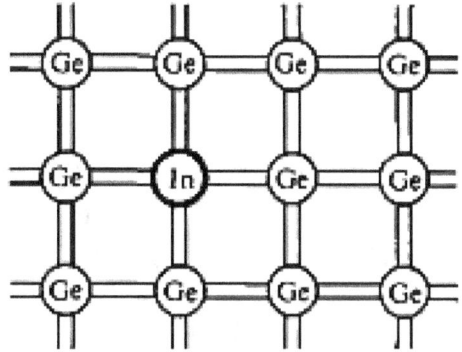

FIGURE 11

Since electron conductivity is the major type of conductivity in crystals with a donor impurity, semiconductors containing this impurity are called electronic, or n–type semiconductors.

In n–type semiconductors at low temperatures, electron conductivity is predominant. At elevated temperatures, say, at room temperature, the conduction band also contains electrons coming from the valence band due to the rupture of valence bonds as well as electrons from the donor level. These transitions are, as we know, accompanied by holes appearing in the valence band and by consequent hole conductivity. Nevertheless, the electron conductivity exceeds the hole conductivity by many times.

For example, if there is only one arsenic atom per 10^7 germanium atoms, in a germanium crystal, the concentration of conduction electrons at room temperature is 2000 times higher than the hole concentration.

Charge carriers whose concentration dominates in a semiconductor under consideration are called majority carriers; charge carriers of the opposite sign are called minority carriers. Naturally, electrons are majority carriers in an electronic semiconductor, while holes are minority carriers.

2.5.2. Hole Semiconductors. Let us now consider the case when a germanium crystal contains a trivalent indium atom (Fig. 21) instead of a pentavalent arsenic atom. The indium atom lacks one electron to create covalent bonds with its four nearest germanium atoms; in other words, in the germanium crystal lattice one double bond is not satisfied. In principle, a saturated covalent bond with the fourth neighbor can be ensured by a transition of an electron from another germanium atom to the indium atom, but, the electron will need some additional energy to do this. Hence, at temperatures close to $T = 0$, when there is no source of

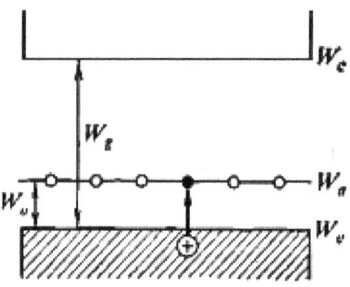

FIGURE 12

this additional energy, valence electrons of the germanium remain with their atoms, and the indium impurity atoms remain neutral with unsatisfied fourth bonds. However, the presence of indium atoms in the crystal makes possible in principle transitions of electrons which have acquired a certain additional energy W_a, to the higher energy levels required to form additional bonds with indium atoms (Fig. 22). Obviously, at $T = 0$ our semiconductor does not conduct electricity because there are no free charge carriers in it (neither: electrons in the conduction band nor holes in the valence band).

As the temperature rises, electrons acquire additional energy of the order of W_a due to thermal lattice vibrations (in the case under consideration, $W_a = 0.01\ eV$) and may go from germanium atoms to indium atoms. A vacancy (hole) is left where the electron moved from. Naturally, a reverse transition is also possible, i.e. the electron may return to the germanium atom. If another valence electron occupies the vacancy while the original electron is at the indium atom, the original electron will have to remain there, thus converting the indium atom to a negatively charged ion bonded with the lattice and hence immobile. The vacancy in the system of valence bonds, formed after the departure of the electron (Fig. 23), thus becomes a free hole. The formation of holes in the valence band (see Fig. 22) signifies that the hole-type conductivity has become possible in the crystal. This type of conductivity determined the name hole, or p—type semiconductors. Impurities introduced into a semiconductor to trap electrons from the valence band are called acceptor impurities, and the energy levels associated with them are called acceptor levels.

2.6. Effect of Temperature on the Charge Carrier Concentration in Semiconductors

When we considered semiconductors with different types of conduction, we emphasized that their conductivity is in principle determined

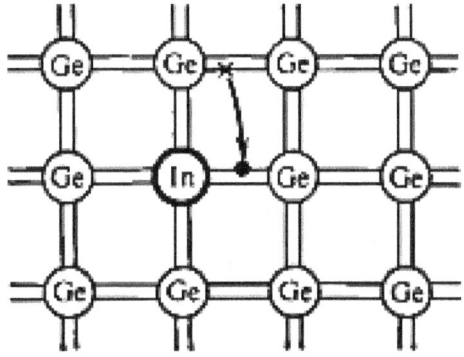

FIGURE 13

by temperature. Semiconductors become capable of conducting electric current only at temperatures considerably higher than absolute zero because it is only then that free charge carriers appear.

Let us now see what difference there is between the effect of temperature on the electric properties of intrinsic and on those of extrinsic semiconductors.

2.6.1. Effect of Temperature on Conductivity of Intrinsic Semiconductors. A peculiar feature of intrinsic semiconductors is that both types of charge carrier-electrons and holes-appear in them simultaneously and in equal amounts. From the moment the energy of thermal lattice vibrations becomes high enough for the electron transfer from the valence band to the conduction band, any further increase in temperature is accompanied by an exceptionally rapid increase in free charge carrier concentration. For example, an increase in temperature from 100 to 600 K in intrinsic germanium raises the carrier concentration by 17 orders of magnitude (10^{17} times).

In order to depict this dependence graphically, a semi-logarithmic scale is normally used. The reciprocal of temperature, $1/T$, is plotted along the x–axis, while the value of $log\ n_i$ (logarithm of concentration) is plotted along the y–axis (Fig. 24). On this scale, the $n_i = f\ (1/T)$ dependence is a straight line intersecting the x–axis at a certain point $1/T$. The charge carrier concentration rises with temperature because electrons from deeper and deeper levels of the valence band move to the conduction band.

If we extrapolated the straight line $n_i = f\ (1/T)$ to the intersection with the y–axis (this would correspond to $T = \infty$), we would obtain the value of $log\ n_{io}$ (where n_{io} is the concentration of the valence electrons in the given semiconductor). It is impossible to attain this state

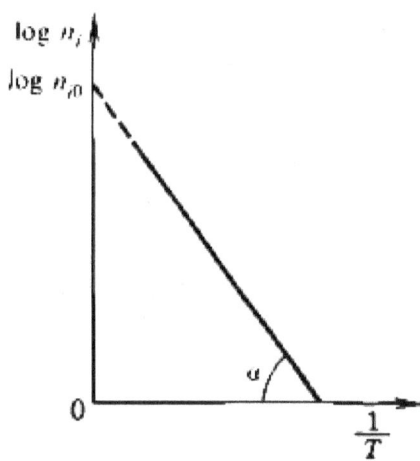

FIGURE 14

experimentally because the crystal lattice would be destroyed and the semiconductor would be melted before all the electrons participating in the formation of covalent bonds travel to the conduction band.

Since the concentration of electrons and holes in an intrinsic semiconductor is the same at any temperature, the $log\ p_i = f\ (1/T))$ dependence will also be the straight line identical to one shown in Fig. 24. Using the equality $n_i = p_i$, we can write

$$n_i p_i = n_i^2$$

And since the concentration of charge carriers in each intrinsic semiconductor only depends on the temperature, it is clear that at any fixed temperature the product $n_i p_i$ is constant (the law of mass action):

$$n_i p_i = n_i^2 = Constant$$

At $T = 300K$ (which is usually taken as room temperature), the value of n_i^2 for silicon is equal to $4.2 \times 10^{21}\ cm^{-6}$ while for germanium $n_i^2 = 6.25 \times 10^{21}\ cm^{-6}$. It is interesting to note that the law of mass action is also valid for impurity semiconductors.

2.6.2. Impurity Semiconductors. The free carrier concentration in electronic and hole semiconductors depends on temperature in the same way, so we can limit ourselves to only n–type semiconductors.

Fig. 25 shows a typical $log\ n_n = f\ (1/T))$ curve for impurity semiconductors. Close to absolute zero, there are no free carriers in the conduction band of an electronic semiconductor, as was the case with intrinsic semiconductors. As the temperature rises, however, the conduction electrons in the electronic semiconductor appear much earlier than in the

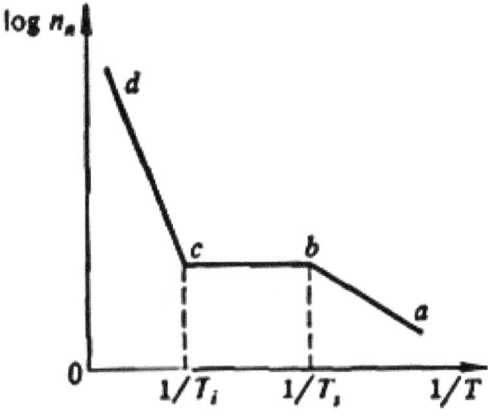

FIGURE 15

intrinsic semiconductor. The reason for this is clear and is because the energy required to detach the fifth valence electron from a donor atom, i.e. to move an electron from the donor level W_d to the conduction band, is almost 100 times less than that needed for the electron transition from the valence band to the conduction band. Recall, for example, that for silicon, $W_d \simeq 0.01\,eV$, while $W_g \simeq 1.2\,eV$. Hence, electrons resulting from transitions labeled 1 in Fig. 26 will be the first to appear in the conduction band. Naturally, there is no difference between these electrons and the electrons arriving to the conduction band from the valence band, but sometimes the former are called impurity electrons to emphasize their origin.

As the temperature rises further, the number of impurity electrons arriving to the conduction band increases rapidly, while the number of electrons remaining on the donor level decreases. In other words, the impurity level is depleted. At a certain temperature T_s (depletion temperature) all the electrons from the donor level are transferred to the conduction band. In this case, the concentration of conduction electrons becomes practically equal to the concentration of the donor impurity N_d (doping level) of the semiconductor:

$$n_n = N_d$$

In most semiconductors impurity levels are depleted at low temperatures. For example, in germanium with a doping level of arsenic atoms $N_d \simeq 10^{16} cm^{-3}$, the temperature of depletion of the donor level is about $30K$. Naturally, the depletion process is dependent on the value of the activation energy W_d of the impurity centers and on their concentration

45

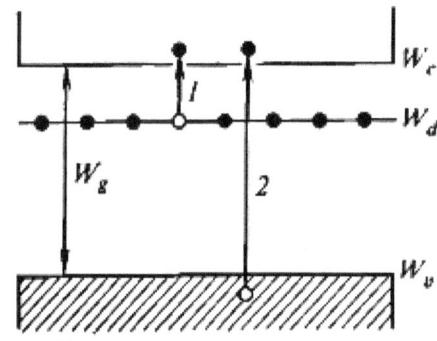

FIGURE 16

N_d. Thus the depletion temperature T_s is the higher, the larger the activation energy and the impurity concentration.

At $T > T_s$ the number of free electrons remains constant over rather a wide temperature range (region bc in Fig. 25). This can be explained by the fact that impurity levels are completely depleted, and the energy of thermal lattice vibrations is still insufficient to excite the valence electrons. In this temperature range, the free carrier concentration can be determined from the relation $n_n \simeq N_d$.

With a further increase in temperature, the process of electron transitions from the valence band to the conduction band becomes more and more intense (these transitions are denoted by 2 in Fig. 26). At a certain temperature T_i, the number of such transitions becomes so large that the concentration of intrinsic electrons, i.e. electrons coming from the valence band, becomes comparable with the concentration of impurity electrons. The temperature T_i is called the temperature of transition to intrinsic conductivity. Any further increase in temperature results in such a rapid growth of intrinsic carrier concentration (region cd in Fig. 25) that we can ignore the impurity electrons and assume that $n_n \simeq n_i$. In this temperature range, an electronic impurity semiconductor does not differ from an intrinsic (pure) semiconductor either in the nature of conductivity (which is now mixed electron and hole, since the departure of electrons from the valence band is associated with a simultaneous appearance of holes) or in the concentration of free charge carriers.

For most impurity semiconductors, the temperature of transition to intrinsic conductivity considerably exceeds room temperature. For example, the T_i for n-type germanium with a doping level of $n_d \simeq 10^{16} cm^{-3}$ is approximately $450K$. The value of T_i depends on the doping level N_d and on the forbidden band width W_g of the semiconductor: the higher

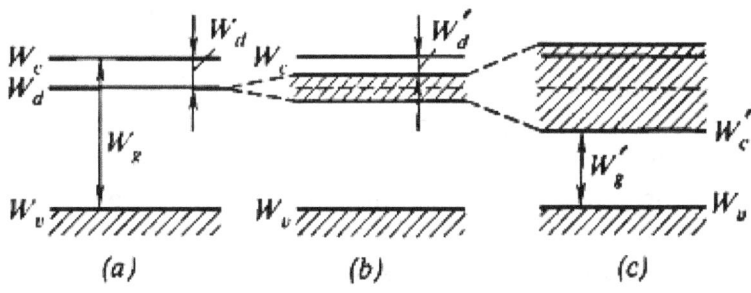

FIGURE 17

the doping level and the wider the forbidden gap, the higher is the temperature of transition to intrinsic conductivity.

2.6.3. Degenerate Semiconductors. Semiconductors with very high doping level ($\sim 10^{18} - 10^{19}\, cm^{-3}$) are especially interesting. They are called degenerate semiconductors, or semi metals. Degenerate semiconductors are very important in semiconductor technology.

It was mentioned above that the addition of a donor impurity to an intrinsic semiconductor leads to a discrete donor level appearing in the energy spectrum of the semiconductor (to be more precise, in its forbidden band) and separated from the bottom of the conduction band by a quite narrow energy interval W_d (Fig. 27 a). At quite low impurity concentrations (the impurity atoms are so far apart that their interaction can be ignored), this donor level can be represented as a line. As the concentration of impurity increases, the impurity atoms become closer and closer to each other, and the orbits of their fifth valence electrons, which are not involved in the lattice formation, start overlapping. These electrons can now move freely from one impurity atom to another. In other words, the valence electrons no longer belong to individual impurity atoms and become common. This means that they can move from one impurity atom to another within the entire crystal.

In the language of the band theory, an increase in the impurity concentration, accompanied by the mutual effects of impurity atoms on each other, leads to the removal of degeneracy in these atoms, the splitting of the donor impurity level into sub levels, and the formation of the impurity band (Fig. 27 b). Essentially, the formation of the energy band from an individual degenerate level in a system of impurity atoms is similar to the band formation from the energy levels of individual atoms when they approach each other to form a crystal (see Fig. 7).

The higher the doping level, the closer are impurity atoms to each other, the stronger their mutual effect and the more the impurity level is smeared. Finally, the impurity band broadens so much that it overlaps with the conduction band (Fig. 27 c) thus forming a hybrid band filled only partially by electrons. Now the impurity electrons which previously occupied the impurity band can travel freely to the conduction band. Since this band contains free higher-lying energy levels, the impurity electrons may acquire an additional energy in an external electric field and hence get accelerated. As was shown above, this indicates the appearance of conductivity in a crystal.

It should be stressed here that the conductivity in degenerate semiconductors is only associated with a high concentration of donor impurity. that leads to overlapping between the impurity band and the conduction band, and is not associated with the temperature of the crystal. In this type of semiconductors, high concentration of conduction electrons can be observed even at absolute zero ($T = 0$). For this reason, they sometimes are called semi metals. The independence of the conduction electron concentration on temperature in degenerate semiconductors is retained up to T_i. At higher temperatures, electron transitions from the valence band to the conduction band become dominating, which causes intrinsic conductivity in the semiconductor.

As the impurity concentration increases, the , broadening of impurity levels into a band is first accompanied by a decrease in the energy of activation of impurity electrons, since the value of W_d' (see Fig. 27 b) is lower than W_d (see Fig. 27 a). This leads to the appearance of impurity electrons at lower temperatures. When the impurity concentration becomes so high that the impurity band and the conduction band overlap, the forbidden band width becomes smaller: W_g' (see Fig. 27 c) is less than W_g (see Fig. 27 a). For example, the forbidden band width in germanium with a doping level of the order of $10^{19}\,cm^{-3}$ decreases from $0.7\,to\,0.5\,eV$.

The $log\,P_p = f(1/T)$ dependence for p—type semiconductors is described by a curve identical to the one shown in Fig. 25. In this case, the regions ab and bc correspond to the hole conductivity and cd, to the intrinsic conductivity. The increase in the hole concentration with temperature in the ab region is due to the trapping of electrons from the valence band by acceptor impurity atoms and the appearance of free holes capable of carrying the charge. When the acceptor level is filled, i.e. when all impurity atoms complete their bonds in the crystal lattice, saturation is attained, characterized by the independence of hole impurity concentration on temperature (bc region). At higher temperatures (*cd* region), a sharp increase in the intrinsic carrier concentration is observed due to the transition of electrons from the valence band to the conduction band.

It should be noted that all impurity semiconductors always contain, in addition to the majority carriers, minority carriers as well, practically at any temperature. Naturally, the number of minority carriers in the low temperature region is small. Concentrations of majority and minority carriers are related through the laws of mass action, which states that the product of concentrations of majority and minority carriers in any semiconductor is equal to the square of the concentration of intrinsic carriers in the corresponding intrinsic semiconductor at the same temperature: $n_p p_p = n_i^2$: and $n_n p_n = n_i^2$:, where n_n and p_n are the concentrations of electrons and holes in electronic semiconductor, while p_p and n_p are the concentrations of holes and electrons in a hole semiconductor. The law of mass action implies that the addition of an active impurity to a semiconductor increases the concentration of the majority carriers and simultaneously decreases the concentration of the minority carriers so that the product of the concentrations of these carriers remains unchanged. For example, if pure germanium with an intrinsic carrier concentration at room temperature $n_i. = 10^{13}\, cm^{-3}$ is doped with such an amount of donor impurity that the number of free electrons increases 1000 times, and the majority carrier concentration becomes $n_n. = 10^{16}\, cm^{-3}$, the number of minority carriers (holes) in this case will decrease 1000 times, and their concentration would become $p_n = 10^{10}\, cm^{-3}$. Thus, the minority carrier concentration will be million times lower than the majority carrier concentration. This decrease in the minority carrier concentration can be explained as follows

As a result of the appearance of the large number of conduction electrons, the lower energy levels in the conduction band are filled. Hence, the electrons from the valence band may move only to the higher-lying, unfilled levels of the conduction band. The energy required for these transitions is higher than in the case of a free conduction band. As a result, the probability of electron transitions from the valence band to the conduction band, and hence the number of holes formed in the valance band decrease.

2.7. Temperature Dependence of Electrical Conductivity of Semiconductors

It was shown above that electrical conductivity is expressed by the formula

$$\sigma = enu$$

where n is the charge carrier concentration that determines the conductivity of a given substance and u the mobility of these carriers. Charge carrier can be either electrons or holes. It is interesting to note that although electrons are known to act as free charge carriers in most metals, this role is played by holes in some metals. Zinc and beryllium are typical examples of metals with the hole conductivity.

In order to determine how the conductivity depends on temperature, we must know how the concentration of free carriers and their mobilities depend on temperature. In metals, the concentration of free charge carriers is independent of temperature. Hence, the temperature variation of the conductivity of metals is completely determined by the temperature dependence of the mobility of carriers. On the contrary, the charge carrier concentration in semiconductors strongly depends on temperature, while temperature variations of the mobility are insignificant. However, in the temperature regions where the carrier concentration is constant (the depletion region and the impurity saturation region), the temperature dependence of conductivity is fully determined by the temperature variations of the mobility of carriers.

Mobility is determined by the scattering of charge carriers by various defects of the crystal lattice, i.e. by the change in the velocity of directional motion of carriers when they interact with various defects. The most important is the interaction between carriers and the ionized atoms of various impurities and the thermal vibrations of the crystal lattice. Scattering processes determined by these interactions have different effects in different temperature regions.

At low temperatures, when the thermal vibrations of atoms are so small that they can be neglected, the scattering by ionized impurity atoms plays the major role. On the other hand, at higher temperatures the atoms at the lattice sites are considerably displaced from their equilibrium position in the crystal due to the thermal vibrations and so thermal scattering predominates.

2.7.1. Scattering by Ionized Impurity Atoms.

In impurity semiconductors, the concentration of impurity atoms is many times higher than the concentration of impurities in metals. Even at quite low temperatures, most of the impurity atoms in a semiconductor are ionized. This is quite natural as the origin of the conductivity of these semiconductors is primarily associated with the ionization of impurities. The scattering of carriers by impurity ions turns out to be much stronger than

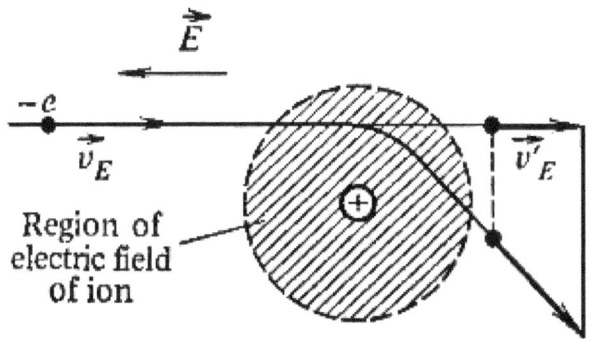

FIGURE 18

the scattering by neutral atoms. This is because for the scattering by ionized atom the carrier need only stray into the electric field created by the ion (Fig. 28), whereas the scattering by a-neutral atom requires a direct collision. When an electron flies through the electric field created by a positive ion, its trajectory bends, as shown in the figure, and the velocity v_E, it acquired in the external field will decrease to v'_E. If the electron comes close enough to the ion its motion after scattering may in general become opposite in the direction to that of the external field

When he was considering the scattering of charged particles by charged centers, Rutherford concluded that the mean free path of particles should be proportional to the fourth power of their velocity:

$$\lambda \propto v_0^4$$

The application of this dependence to the scattering of carriers in semiconductors has led to a very interesting and at first sight unexpected result: at low temperatures the mobility of carriers must increase with temperature. In reality the mobility of carriers turns out to be proportional to the third power of their velocity:

$$u \propto \frac{\lambda}{v_0} \propto \frac{v_0^4}{v_0} = v_0^3$$

On the other hand, the mean kinetic energy of charge carriers in semiconductors is proportional to temperature, $W_k \propto kT$, and hence the mean thermal velocity is proportional to the square root of T (since $W_k = mv^2/2$). Consequently, we get the following relationship between the mobility of carriers and the temperature:

$$u \propto v_0^3 \propto (\sqrt{T})^3 = T^{\frac{3}{2}}$$

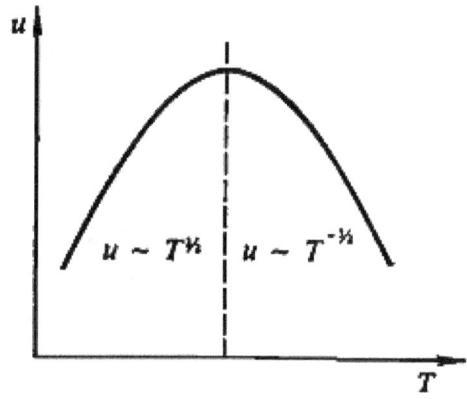

FIGURE 19

At low temperatures where the scattering by ionized impurity atoms dominates and the thermal vibrations of the lattice atoms can be ignored, the mobility of carriers increases with temperature in proportion to $T^{\frac{3}{2}}$ (the left branch of the $u(T)$ curve in Fig. 29). Qualitatively, this dependence can be easily explained because the higher the thermal velocity of carriers, the shorter is the time they stay in the field of an ionized atom and hence the less their trajectory is distorted. Thus their mean free path and mobility increase.

2.7.2. Scattering by Thermal Vibrations. As the temperature increases, the mean velocity of the thermal motion of carriers increases to such an extent that the probability of their scattering by ionized impurity atoms becomes very low. At the same time, the amplitude of thermal vibrations of the lattice atoms increases so that the scattering of carriers by thermal vibrations becomes predominant. The mean free path of carriers, and hence their mobility, decrease as the material is heated further due to the intensification of the scattering by thermal vibrations.

The shape of the $u(T)$ curve in the high-temperature region is different for different semiconductors. It depends on the nature of the semiconductor, the width of the forbidden band, doping level, and some other factors. For typical covalent semiconductors, germanium and silicon in particular, the $u(T)$ dependence has the form

$$u \propto T^{-3/2}$$

for moderate doping levels (see the right branch in Fig. 29)

Thus, the mobility of carriers in semiconductors grows in proportion to $T^{3/2}$ at low temperatures and decreases in inverse proportion to $T^{3/2}$ at high temperatures.

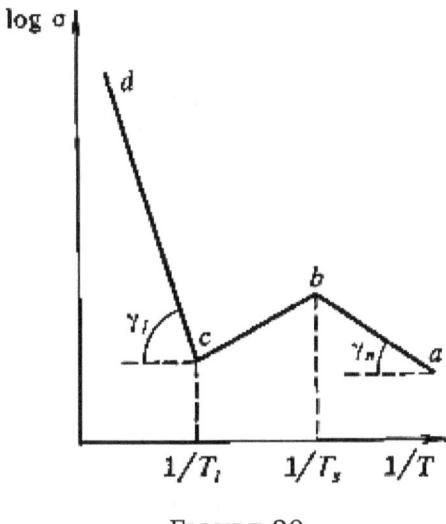

FIGURE 20

2.7.3. Temperature Dependence of Conductivity of Semiconductors. If we know how the mobility and carrier concentration depend on the temperature in a semiconductor, we can establish how its conductivity depends on temperature. The $\log \sigma = f(1/T)$ curve is shown schematically in Fig. 30. It can be seen that the shape of this curve is similar to that for the $\log n = f(1/T)$ curve shown in Fig. 25. Since the temperature dependence of the carrier concentration is much stronger than the temperature dependence of the mobility, the temperature dependence of the conductivity $\sigma(T)$ in the regions of impurity

conductivity (region ab) and intrinsic conductivity (region cd) is almost completely determined by the temperature dependence of the carrier concentration. The slopes of these sections of the graph depend on the ionization energies of the donor impurity atoms and on the forbidden band width of the semiconductor. The slope γ_n is proportional to the energy of detachment of the fifth valence electron from the atom of the donor impurity. Therefore, if we can obtain experimentally a graph describing the temperature variation of the conductivity of a semiconductor in the impurity region ab, we can find the value of the activation energy for the donor level, i.e. the separation W_d between the donor level and the bottom of the conduction band (see Fig. 20). The slope γ_i is proportional to the energy required for an electron to go from the valence band to the conduction band, viz. the energy required to create intrinsic carriers in a semiconductor. Consequently, having determined experimentally the temperature dependence oi the conductivity of a semiconductor in the intrinsic region cd, we can find the forbidden band width W_g (see Fig.

27). The values of W_a and W_g are the most important characteristics of a semiconductor.

The difference between the $\sigma(T)$ and $n(T)$ dependencies is mainly observed in the bc region which lies between the impurity depletion temperature T_s and the temperature T_i of the transition to intrinsic conductivity. This region corresponds to the ionized state of all the impurity atoms, while the energy of the thermal vibrations is still insufficient for creating intrinsic conductivity. For this reason, the carrier concentration (which is practically equal to the concentration of impurity atoms) does not change with increasing temperature. In this region, the temperature dependence of conductivity is determined by the temperature dependence of the carrier mobility. At a moderate impurity concentration, the scattering by thermal lattice vibrations is the main scattering mechanism in this temperature region for most of semiconductors. This mechanism is responsible for a decrease in the mobility of carriers, and hence in the conductivity of semiconductors, as the temperature increases within bc.

For degenerate semiconductors, the scattering of carriers by ionized impurity atoms remains the same up to high temperatures due to the high doping level which causes the electric fields of ions to overlap. As we mentioned above, this mechanism is characterized by an increase in the carrier mobility with temperature.

CHAPTER 3

Non equilibrium Processes in Semiconductors

3.1. Generation and Recombination of Non equilibrium Charge Carriers

3.1.1. Free Carrier Generation. It was shown that at a temperature above absolute zero, all semiconductors have some free carriers. Generally, the free carrier concentration rises with temperature. The appearance of free carriers is explained by transitions of electrons from the valence band or from the donor levels (to the conduction band. This process is called the generation of free carriers (Fig. 31)

3.1.2. Free Carrier Recombination. If, however, the generation of carriers were the only process in a semiconductor, more and more electrons would arrive to the conduction band, and in a certain time all the valence electrons would be found in the conduction band. Nothing of this kind is observed experimentally. This is because the generation of carriers is accompanied by the recombination of free carriers. After a certain time, the electrons from the conduction band return to the valence band or to vacancies in the impurity level (Fig. 32). In other words,

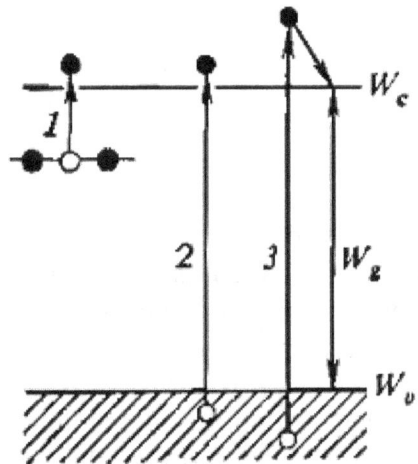

FIGURE 1

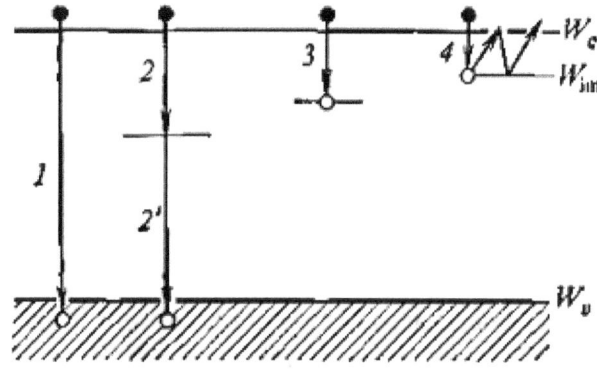

FIGURE 2

after a trip in the interstitial space of the crystal, a free electron sooner or later meets a vacancy and occupies it. If this vacancy is a positively charged impurity ion, then one conduction electron vanishes, but if the vacancy is a ruptured bond in the lattice (hole), two free carriers, viz. the electron and the hole, vanish simultaneously.

3.1.3. Equilibrium Carriers. Thus, the content of free carriers in a semiconductor is determined by ,two opposite and simultaneously occurring processes, viz, generation and recombination. Thermal processes in solids are always inertial, i.e, they proceed quite slowly, hence thermal generation and recombination have time to attain equilibrium at practically any temperature. Consequently, a carrier concentration determined only by the thermal processes is called an equilibrium concentration, and the carriers themselves are called equilibrium carriers. Since the generation and recombination of equilibrium carriers are always balanced (a new pair of charge carriers is immediately created as the result of thermal excitation in place of each recombining pair), the generation and recombination processes are ignored in thermal equilibrium.

3.1.4. Non equilibrium Carriers. Along with thermal excitation, some other factors create free carriers in semiconductors, for example, irradiation by light or bombardment by particles causing ionization. Free carriers may also appear through contact with another body. Free carriers appearing in a semiconductor due to these factors are excess carriers with respect to the equilibrium carriers and are called nonequilibrium carriers.

The mechanism of generation of non equilibrium carriers differs in principle from that of equilibrium carriers. During thermal generation, the heat supplied from outside increases the energy of the thermal vibrations of atoms in the crystal lattice. When this energy becomes high enough for the covalent bonds to rupture, the atoms are ionized and equilibrium carriers thus appear. Consequently, the heat supplied creates free carriers through a mediator, viz. the crystal lattice of the semiconductor. For this reason, the concentration of equilibrium charge carriers is the same throughout the volume of the semiconductor.

On the other hand, when non equilibrium carriers are generated, the energy of the external source of excitation (the energy of photons, the energy of bombarding electrons or other particles) is supplied directly to the valence electrons, while the energy of the crystal lattice remains practically unchanged. Therefore the non equilibrium carriers are not in thermal equilibrium with the crystal when they are formed and their density distribution throughout the volume can be very nonuniform (usually, the concentration of non equilibrium carriers decreases away from the crystal surface or from the exposed region)

As a result of the direct energy transfer to a charge carrier, the electron may acquire an energy which is higher than that required to overcome the forbidden gap. In this case, excess electrons are transferred to the higher energy levels of the conduction band (transition 3 in Fig. 31), the kinetic energy of such non equilibrium carriers being much higher than the mean energy $3kT/2$ which corresponds to the energy of the thermal equilibrium electrons at the bottom of the conduction band. This excess kinetic energy dissipates over time during collisions between the non equilibrium electrons and crystal lattice defects. During $10^{-10} sec$, fast electrons undergo about one thousand scattering acts as a result of which their kinetic energy levels out to that of the thermal electrons. We can assume that the kinetic energy of the non equilibrium carriers becomes comparable to the energy of the equilibrium carries almost immediately after their generation, and hence their mobility is the same as the mobility of the equilibrium carriers. Therefore, the change in conductivity during the generation of non equilibrium carriers is only due to a change in the total carrier concentration. The instantaneous differences in the energy and mobility of the non equilibrium carriers that occur immediately after they are generated are insignificant.

3.1.5. Electra-neutrality Condition. If we denote by Δn and Δp the concentration of excess electrons and holes respectively, the total carrier concentration will be

$$n = n_o + \Delta n$$

$$p = p_o + \Delta p$$

where n_o and p_o are the equilibrium concentrations of electrons and holes. If only intrinsic carriers are generated by the external action on the semiconductor (i.e. transitions from the valence band to the conduction band are excited), and the semiconductor itself contains no volume charge, the concentration of the non equilibrium electrons must be equal to that of the excess holes

$$\Delta n = \Delta p$$

This is the electromagnetically condition..

It was mentioned above that the mobility of the excess carriers is practically the same as the mobility of equilibrium carriers. Hence, we can write the following expression for the additional conductivity $\Delta \sigma$ appearing in a semiconductor under the effect of an external source of excitation:

$$\Delta \sigma = e(u_n \cdot \Delta n + u_p \cdot \Delta p)$$

and for the total conductivity,

$$\sigma = e(u_n(n_o + \Delta n) + u_p(p_o + \Delta p))$$

After the excitation source has been switched off, the semiconductor gradually returns to its equilibrium state. In this case, the free carrier concentration in it again acquires the equilibrium value due to recombination. However,, before the recombination occurs (i.e. before an electron meets a hole), each carrier remains in the free state for a certain time. This time duration depends on many factors and varies between 10^2 and 10^{-9} sec. In order to characterize non equilibrium carriers as a whole, we introduce the mean lifetime τ_n for the non equilibrium electrons and τ_p for the equilibrium holes. In practice the mean lifetime of an electron, for example, is defined as the mean time from the moment of generation of the excess electron to the moment when it is trapped by a hole.

3.1.6. Recombination Rate. The concept of the mean lifetime is closely related to that of the recombination rate (i.e. the rate at which the excess carrier concentration changes). The recombination rate is defined as the nonmember of recombining pairs of carriers per unit time, $\Delta n / \tau_n$ (where Δn is the non-equilibrium carrier concentration at a given instant t). Clearly, the recombination rate is the higher, the greater the non-equilibrium carrier concentration. The recombination rate decreases with time, because the number of excess carriers decreases due to recombination.

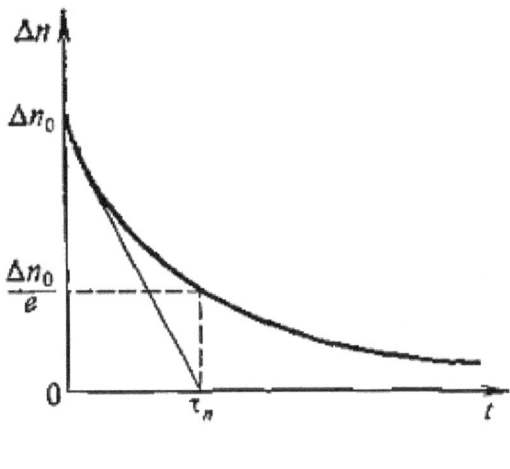

FIGURE 3

Simple calculations show that the non equilibrium carrier concentration decreases exponentially with time:

$$\Delta n = \Delta n_o \, e^{-t/\tau_n}$$

where Δn_o is the initial excess carrier concentration at the instant $t = 0$ (the moment the excitation source is switched off), $e = 2.7$ is the natural logarithmic base.

It follows from this formula that the non equilibrium carrier concentration has decreased e times in a time τ_n, after switching off the source of excitation. Indeed at $t = \tau_n$, the concentration An becomes equal to $\Delta n_0/e$. The Δn vs. t dependence is shown in Fig. 33. This curve can be used to determine the value of τ_n as a segment intercepted on the time axis by the tangent to the curve at the point $t = 0$.

3.1.7. The Concept of Trapping Cross Section. Recombination implies the trapping of an electron by a hole. A hole's ability of trapping an electron is characterized by the so-called trapping cross section. Let us draw through the hole a plane perpendicular to the direction of the motion of the electron approaching the hole. If the electron crosses this plane far from the hole, it will not be trapped but will only change the direction of its motion to a certain extent. If, however, the electron crosses this plane close to the hole, i.e. within a circle of a sufficiently small radius, it is trapped by the hole. The area of this circle is called the effective trapping cross section, or simply the trapping cross section. The trapping cross section strongly depends on the electron velocity relative to the hole and the origin of this hole.

3.1.8. Types of Recombination. There are two types of charge carriers. In the first case, an electron directly moves from the conduction band to the valence band (transition 1 in Fig. 32). This is called inter band (or band-to-band) recombination. In the second case, the electron transition 1 from the conduction band to the valence band occurs in two stages: first, the electron goes from the conduction band to a certain intermediate level (transition 2) created in the crystal by an impurity atom, and then it goes from this level to a vacancy in the valence band (transition 2′). This process is called the recombination through an impurity site. In both cases, the transition of the electron from the conduction band to the valence band is accompanied by the liberation of energy approximately equal to the forbidden band width W_g (as a free, electrons at the bottom of the conduction band and holes at the top of the valence band take part in recombination).

The energy liberated during recombination can be either emitted in the form of a radiation quantum or transformed to the energy of the crystal lattice. Accordingly, we have either radiative or non radiative recombination.

3.1.9. Radiative Recombination. In this process, the energy of an electron combining with a hole is transformed into the energy hv of a photon. The act of recombination in this case is similar to the return of an excited atom to the ground state (the electron goes over from the excitation orbit to the ground-state orbit). According to the law of the conservation of energy, the energy of a photon emitted during the inter band recombination is given by

$$hv = W_c - W_v = W_g$$

where W_c is the electron energy before recombination (at the bottom of the conduction band) and W_v the electron energy after recombination (at the top of the valence band).

In the majority of cases, radiative recombination is associated with direct band-to-band electron transitions. However, these transitions can occur only in very pure semiconducting crystals with a narrow forbidden gap, e.g. indium antimonide (InSb) crystals for which the forbidden band width W_g is only $0.18 eV$. The radiative recombination lifetime of an electron in these crystals is short (of the order of $10^{-7} sec$). Hence, practically all the electrons recombine, emitting a photon.

As the forbidden band width increases, the probability of radiative recombination becomes lower. For example, in intrinsic germanium, for which $W_g = 0.72 eV$ at room temperature, the electron lifetime relative to radiative recombination is 1 sec, while the lifetime relative to non radiative recombination equals 10^{-3} to 10^{-4} sec. This means that one recombination accompanied by the emission of a photon occurs per several thousand non radiative recombination transitions.

3.1.10. Non radiative Recombination. This form of recombination, which is typical of semiconductors with a sufficiently wide forbidden gap, involves, as a rule, impurity sites. The energy liberated during the transition of an electron from the conduction band to the valence band is rather high and cannot be transferred to the crystal lattice at once. The probability of this act is as low as the probability of a simultaneous collision of ten particles at one point. On the other hand, in the case of recombination through impurity levels, the electron energy is transferred to the crystal lattice in two stages. In this case, the closer the energy level of the impurity atom to the center of the forbidden gap, the higher the probability that the recombination proceeds through this atom, since with such an arrangement the energy is liberated in two equal portions both equal to $W_g/2$. Band theory would say that during the first stage the impurity site traps an electron and during the second, a hole. In principle, these stages can be passed in the reverse order. Their sequence is determined by the probabilities of trapping an electron and a hole by the impurity site. If the energy level of the impurity site lies closer to the bottom of the conduction band, the probability of trapping an electron will be higher than the probability of trapping a hole, because the lower energy will be liberated during the electron trapping (in general, the lower the liberated energy, the higher the probability of the process). In this case, after having been trapped by the impurity site, an electron has to "wait" for a hole for some time. As a matter of fact, although there are many holes, the electron cannot recombine with any of them. The recombination is possible only when the electron can give the liberated energy to the crystal lattice.

3.1.11. Trapping Sites and Trapping Centers. It may happen that the recombination of excess carriers has already been completed but the electron at the impurity center is still awaiting the required hole. Such a situation usually occurs when the level of the impurity center is rather far from the middle of the forbidden gap (transition 3 in Fig. 32). In this case, the impurity center is called the electron trapping site.

If the level of an impurity center lies near the bottom of the conduction band, so that the energy difference $W - W_{im}$, becomes comparable with the energy of thermal lattice vibrations, an electron trapped by such a site may again be thrown back to the conduction band (case 4 in Fig. 32). This process may be repeated many times before the electron "drops" down to the valence band. Such impurity centers are called trapping centers. The presence of these centers may considerably increase the mean lifetime of non equilibrium carriers.

In high-frequency semiconductor devices it is required that excess carriers rapidly vanish after the excitation signal has been switched off. For this reason, gold, nickel, copper, and some other impurities are used to increase the recombination rate in devices based on germanium or silicon single crystals. These impurities form recombination levels near the middle of the forbidden gap.

3.1.12. Surface Recombination. Free charges may disappear not only as a result of recombination in the bulk but also due to surface recombination which proceeds in many cases much more rapidly than recombination in (the bulk. This can be explained by the fact that there are always many adsorbed impurity atoms and various defects on the surface of a semiconductor, which are effective recombination centers. Surface recombination plays an especially important role in thin samples where the volume is small and the surface is relatively large.

Since recombination on the surface is more intense than that in the bulk, the number of free carriers near the surface is less than in the bulk of a semiconductor. This difference in concentration causes a flux of free carriers from inside to the surface. The greater the difference in the concentrations (i.e. the more intense the surface recombination), the higher the velocity of carriers moving to the surface. In order to suppress surface recombination, semiconductor samples are specially treated (chemical etching, thorough washing, etc.) to remove recombination centers from the surface. An appropriate treatment of the surface can decrease the rate of surface recombination by two orders of magnitude.

3.2. Diffusion Phenomena in Semiconductors

Regardless of the way the non equilibrium carriers were created (the introduction of excess carriers through the interface with a metal or their generation by irradiation or bombardment by fast particles), the distribution of these carriers over the semiconductor is always nonuniform. It is quite clear that the concentration of non equilibrium carriers near the semiconductor surface where they are created is much higher than in the bulk. The difference in concentrations leads to a diffusion of the non equilibrium carriers from the region with the higher concentration to the region with the lower concentration.

3.2.1. Diffusion Current. The diffusion flux of non-equilibrium charge carriers creates an electric current called the diffusion current.

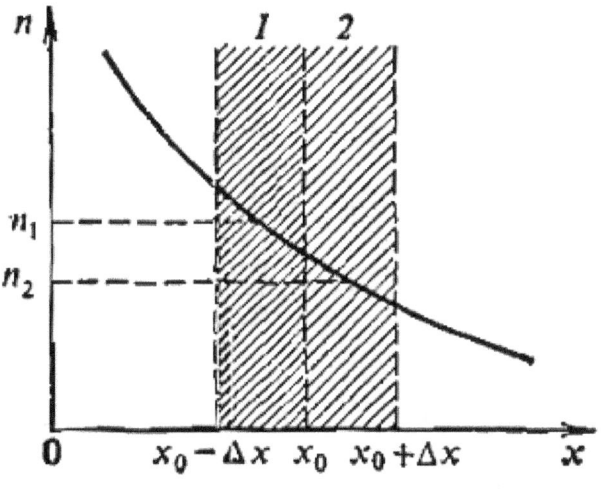

FIGURE 4

The diffusion current can be either electron or hole in nature. Let us consider some of the features and regularities of the diffusion of the non equilibrium electrons.

Suppose the electron concentration in a semiconductor decreases, as shown by the curve in Fig. 34, as we move from the outer surface (corresponding to $x = 0$) to the bulk. Now let us cut the semiconductor by an imaginary plane that passes through the point $x = x_0$ and is perpendicular to the x−axis, and isolate two adjacent layers of thickness Δx. AS a result of random thermal motion, all the electrons contained in layer 1 at an initial moment leave it after some time. Since the electrons can move left or right with equal probability, we can assume that half of electrons from layer 1 will cross the plane $x = x_0$. At the same time, half the electrons from layer 2 (which has the same thickness Δx) will cross this plane in the opposite direction. As the average electron concentration n_1 in layer 1 is higher than the average electron concentration n_2 in layer 2, the number of electrons crossing the boundary between the layers from left to right will be larger than the number of electrons crossing it in the opposite direction. The difference between these fluxes is equal to the resultant diffusion flux which determines the appearance of the diffusion current.

The intensity of the diffusion flux is determined by the difference in the electron concentrations of the layers in contact. In turn, the difference between n_1 and n_2 is determined by the change in the electron concentration per unit length in the direction perpendicular to the boundary between the layers ($\Delta n/\Delta x$). The density in of the electron diffusion current can be expressed by the formula :

$$i_n = eD_n \frac{\Delta n}{\Delta x}$$

where D_n is the electron diffusion coefficient.

For p–type semiconductors, the density of current due to the motion of holes is expressed by a similar formula:

$$i_p = eD_p \frac{\Delta p}{\Delta x}$$

where D_p is the electron diffusion coefficient.

3.2.2. The Einstein Relation. The diffusion coefficient D depends on the nature and structure of a material. It is proportional to the carrier mobility u and the absolute temperature T of the crystal :

$$D = \frac{kT}{e} u$$

This is the Einstein relation.

Since the mobility of electrons is greater than the mobility of holes, the diffusion coefficient for electrons is always lighter than that for holes. For example, in germanium at room temperature, the electron diffusion coefficient is twice the hole diffusion coefficient, and in silicon, it is three times greater.

3.2.3. Holes Pursue Electrons. It is interesting to note that in a homogeneous semiconductor the diffusion of one type of carriers does not practically lead to a violation of the electro neutrality condition: if we measure the excess carrier concentration at any point, it always turns out that $\Delta n = \Delta p$. This is because the diffusion of one type of carriers is always accompanied by a simultaneous transfer of carriers of the opposite sign. Let us consider a specific example

Suppose that some electrons leave a certain volume of a semiconductor. Naturally, this immediately violates the electro neutrality: Δp becomes larger than Δn, and a volume charge $e(\Delta p - \Delta n)$ appears in the region. This charge creates an electric field directed to the region where the electrons went. The intensity of this field can be very high; in ordinary semiconductors it attains the values $10^5 V/cm$ or higher. Such a strong field causes an intense flux of holes to leave the volume and thus restore the upset balance. Thus, the volume charge soon disappears. Moreover, the holes catching up with the electrons prevent electro neutrality from being disturbed in new regions. Of course, any abruptly violated electro neutrality cannot be restored immediately, requiring a certain time τ_0, called the dielectric relaxation time, to occur. However, this period is so small (in normal conditions, $\tau_0 \simeq 10^{-12}$ sec) that a diffusion flux of one type carriers is actually accompanied by a parallel transfer of carriers of the opposite sign, owing to which the electro neutrality condition remains valid for any volume of a uniform semiconductor.

3.2.4. Diffusion and Recombination. The diffusion and recombination of non equilibrium carriers are closely related. We shall illustrate this by considering the propagation of non equilibrium holes in an electronic semiconductor. Suppose that there is a source of holes on one surfaces of an n—type semiconductor that creates on this face (where $x = 0$) a certain excess concentration $\triangle p$, of non equilibrium holes. Since the concentration of holes in the bulk of the semiconductor is lower than at the surface, holes diffuse into the semiconductor, i.e. the diffusion current appears. If there were no carrier recombination in the bulk of the semiconductor, the excess holes would reach in a certain time the opposite face of the sample, and the constant concentration $\triangle p$, of excess carriers would be established throughout the semiconductor. However, this is not observed in reality because after diffusing from the surface, the non equilibrium holes recombine with electrons which move to the surface layer from inside, and their concentration decreases in the direction from the surface to the bulk of the sample.

The decrease in the excess hole concentration away from the semiconductor surface on which they were created is expressed by the exponential law (Fig. 35)

$$\triangle p = \triangle p_0 e^{-x/L_p}$$

The parameter L (Lp in the case of holes) is called the diffusion length for carriers. The value of L_p is equal to the distance over which the excess hole concentration decreases e times.

The relation between the diffusion and recombination processes is described by the following

$$L_p = \sqrt{D_p \tau_p}$$

where τ (τ_p for holes) is the mean lifetime of non equilibrium carriers (recall that the mean lifetime of non equilibrium carriers is the mean time between their generation and recombination). The diffusion length L is the average distance covered by non equilibrium carriers during their lifetime, i.e. the distance between their generation and recombination sites.

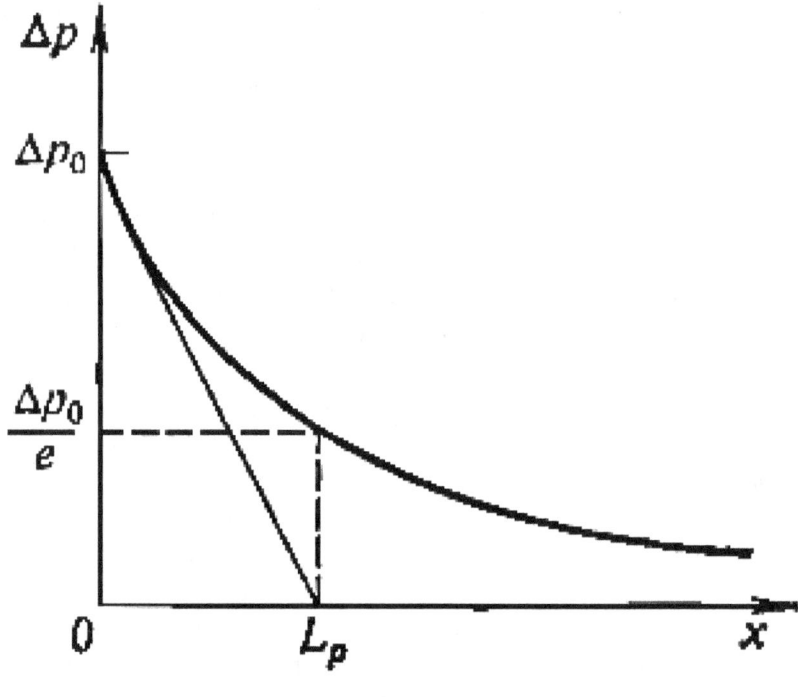

FIGURE 5

3.3. Photo-conduction and Absorption of Light

Photo-conduction is one of non equilibrium processes that occurs in semiconductors. It consists in the appearance of a change in the conductivity of a semiconductor under the effect of some radiation (infrared, visible, or ultraviolet). As a rule, when a semiconductor is irradiated its conductivity $\sigma - neu$ increases owing to an increase in the free carrier concentration n (the mobility u of the non equilibrium carriers is practically the same as the mobility of equilibrium carriers).

3.3.1. Photo-induced Processes. There are three main causes of generation of excess mobile carriers under the action of light (Fig. 36):

(1) the light quanta interacting with electrons in the donor impurity levels transfer them to the conduction band, thus increasing the concentration of conduction electrons (*case* 1);

(2) the light quanta excite the electrons in the valence band and transfer them to the acceptor levels, thus creating free holes in the valence band and increasing the hole conductivity of the semiconductor (*case* 2);

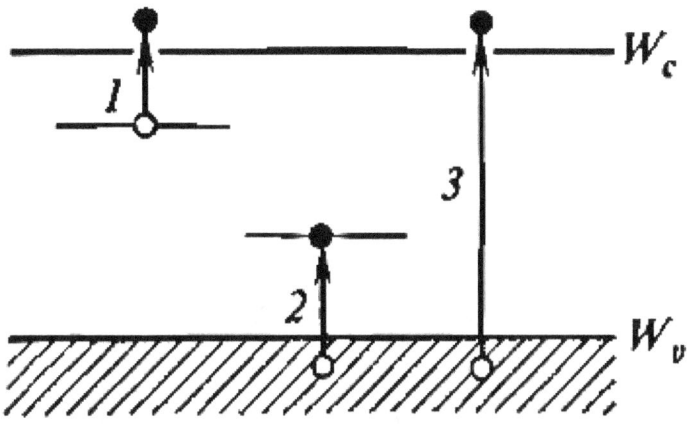

FIGURE 6

(3) the light quanta directly transfer the electrons from the valence band to the conduction band, thus creating simultaneously both mobile holes and free electrons (*case*3).

In the first two cases every effective interaction between a quantum and an electron is accompanied by the appearance of one free carrier (either an electron or a hole), while in the third case each interaction results in the formation of two mobile charge carriers.

3.3.2. Laws Governing Photo-conductivity. Let us consider the variation of conductivity of a semiconductor under the action of light by using as an example the electron photo-conduction created during the excitation of impurity levels. If we assume that after the source of light is switched on, g free electrons are generated every second in a unit volume of the semiconductor, at first the number of free conduction electrons will increase very rapidly, but then the rate of this increase will gradually fall. This is explained by the accelerating process of recombination (the recombination rate is proportional to the number of excess carriers; hence, the higher the concentration of free electrons and the greater the number of vacancies formed in the donor levels, the more intense the recombination). In the long run, the generation of electrons and their recombination (to be more precise, their return to vacant donor levels) balance each other, and a stationary concentration Δn_{st} of excess photo-electrons will be established:

$$\Delta n_{st} = g\tau_n$$

where τ_n is the lifetime of photo-electrons.

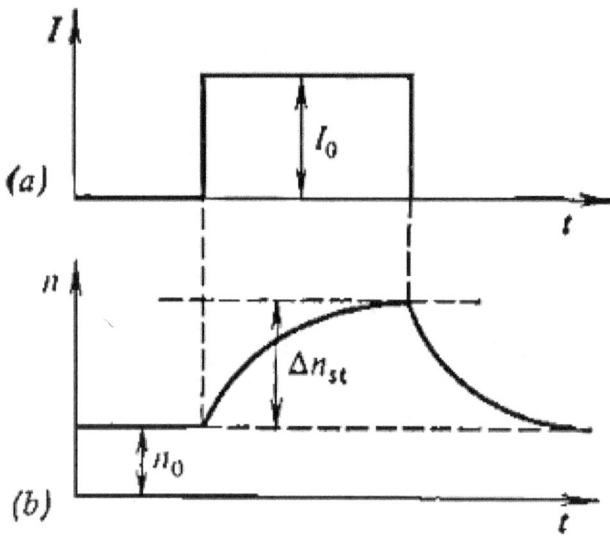

FIGURE 7

The increase in the concentration of electrons generated by the light is expressed by the following formula:

$$\Delta n = \Delta n_{st}(1 - e^{-t/\tau_n})$$

As time passes, the concentration Δn asymptotically tends to Δn_{st} as is shown in Fig. 37b (Fig. 37 a shows the change in the intensity I of light falling on the semiconductor surface).

When the light is switched off, the generation of electrons ceases, and as a result of recombination, the concentration Δn drops exponentially to zero:

$$\Delta n = \Delta n_{st} e^{-t/\tau_n}$$

The semiconductor thus returns to the initial state characterized by the equilibrium electron concentration no. The rate of the increase (as well as the rate of the decrease) in the photo-electron concentration is completely determined by the value of τ_n, which depends on the properties of a specific sample.

Since only the concentration of photo-electrons changes upon irradiation, while their mobility remains the same, the photo-conductivity of a semiconductor can be expressed as follows:

$$\Delta \sigma = e u_n \Delta n$$

The variation of photo-conductivity with time is almost completely identical to the time dependence of the photo-electron concentration shown in Fig. 37. The value of the stationary photo-conductivity is determined by the formula:

$$\Delta\sigma_{st} = eu_n \Delta n_{st} = eu_n g \tau_n$$

which indicates that the magnitude of photo-conductivity is directly proportional to the lifetime of photo-electrons.

3.3.3. Quantum Efficiency. The intensity of light exciting a semiconductor is usually expressed by the number of photons impinging on the sample. Some of the photons are reflected by the semiconductor's surface, some may pass . through the sample, but the remaining photons are absorbed in the bulk. The ratio of the number of photo-electrons g released upon absorption of light to the total number N of absorbed photons is called the quantum efficiency

$$\eta = \frac{g}{n}$$

If each act of photon absorption were accompanied by the generation of a free electron, the quantum efficiency would be equal to unity. However, besides photo-induced processes, there are very many other processes of photon absorption which do not generate photo-electrons and so the quantum efficiency is usually less than unity. Sometimes, the quantum efficiency may exceed unity. This is possible if the energy of a single photon is sufficient for creating two or more photo-electrons (irradiation of a semiconductor by ultraviolet light or X-rays). The situation when $\eta > 1$ can be realized in an intrinsic semiconductor in the following way.

If the energy of an absorbed photon is more than double forbidden band width, the electron receiving this energy and transferred to the conduction band will have a high kinetic energy whose value is determined by the energy difference between the level to which the electron has come and the bottom of the conduction band (Fig. 38). During its trip through the crystal's interstitial space, the electron may give off some of the acquired energy to the crystal lattice by being scattered by various in-homogeneity and impurities (region *AB*) . But if it encounters a semiconductor atom (point *B* on the diagram), it may donate all (or a part) of its excess energy to this atom by detaching an electron from it and thus creating another pair of free charge carriers. Hence, a single photon has created four free carriers (two electrons and (two holes) as

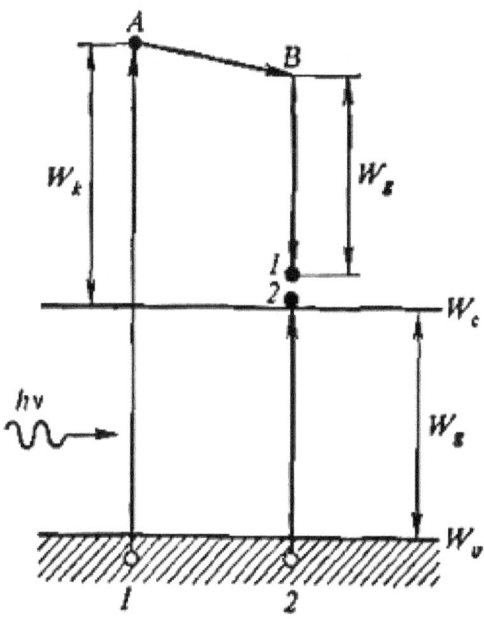

FIGURE 8

a result of two successive acts of ionization. A photon with still greater energy can create even more pairs of charge carriers.

Similarly, non-equilibrium conductivity can be created by bombarding the semiconductor with fast charged particles, e.g. electrons. The primary electrons generate fast secondary electrons in the bulk of the semiconductor, and these in turn create new pairs of carriers by ionizing the atoms of the crystal lattice. This process of free carrier multiplication continues till the energy of the conduction electrons created upon ionization is sufficient to transfer an electron from the valence band to the conduction band.

The quantum efficiency can also exceed unity if the primary photon (quantum of light) that has at least double the energy of the forbidden band width ($h\nu_1 > 2W_g$) does not lose all its energy upon ionization when it enters a semiconductor but only loses that amount required to transfer an electron from the valence band to the conduction band (Fig. 39). As a result, a new photon appears with an energy $h\nu_2$ that is still higher than the forbidden band width. Such a photon can create one more pair of carriers by colliding with an atom of the semiconductor.

3.3.4. Photo-current Spectral Distribution Curve. When the semiconductor sample is connected to an electric circuit and illuminated, then in addition to the dark current (i.e. the current flowing in the sample

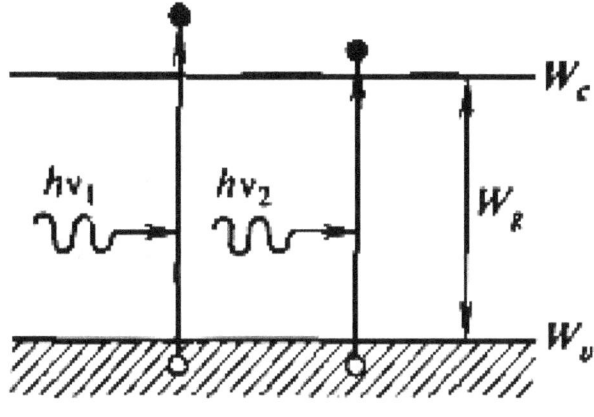

FIGURE 9

without illumination), the photocurrent I_{ph} appears in it. The photo-current intensity depends on the wavelength of light illuminating the sample. At certain values of the wavelength λ the photo-current has maxima, while in some wavelength regions it is equal to zero. The dependence of the photo-current intensity on the wavelength λ of excitation light is called the spectral distribution curve for photo-current (Fig. 40a).

3.3.5. Intrinsic Photo-conductivity. The spectral distribution curve shown in Fig.40a corresponds to an impurity semiconductor. Two photo-current maxima are clearly seen on it and correspond to intrinsic (*region*1) and impurity (*region*2) charge carriers appearing in the semiconductor under the action of light. If a semiconductor If a semiconductor contains no impurity atoms (an intrinsic semiconductor), only maximum 1 will be observed on the photo-current spectral distribution curve. In this case, the absence of photo-conductivity at $\lambda > \lambda_{thr}$, is explained by the fact that in the long-wave range, the photon energy is insufficient for transferring electrons from the valence band to the conduction band. This will be observed for the entire long-wave range where $h\nu \ll W_g$. At a wavelength corresponding to the photon energy comparable with the forbidden band width ($h_\nu \simeq W_g$), an electron near the top of the valence band can go to the conduction band at the expense of the energy of absorbed photon. Since this opportunity appears for a large number of intrinsic electrons, and a pair of carriers (an electron and a hole) is created in each interaction between a photon and an atom, the photo-current increases very rapidly.

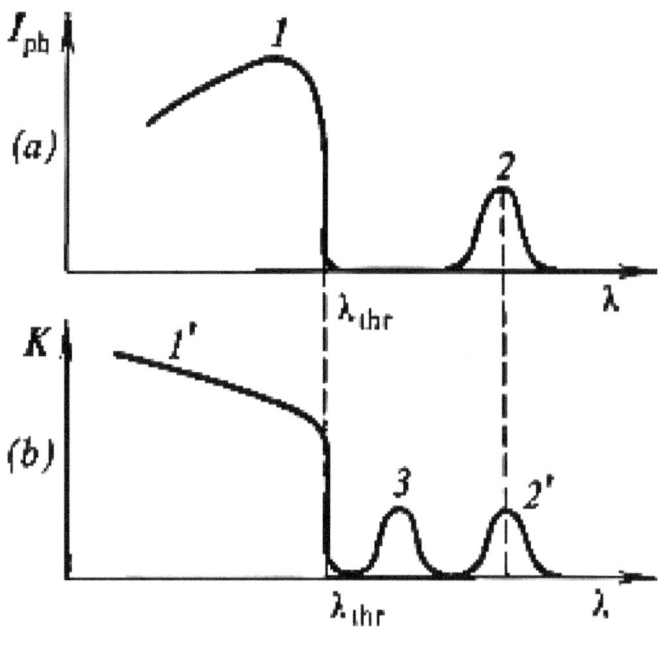

FIGURE 10

The wavelength λ_{thr} at which photo-conductivity appears can be determined from the formula

$$h\nu_{thr} = W_g$$

which defines the position of the photoelectric threshold.

As the wavelength decreases to $\lambda < \lambda_{thr}$ the photo-conductivity of the semiconductor decreases. This can be explained as follows.

When $h\nu \gg W_g$, the light is absorbed so intensely that practically all the incident light is absorbed in a very thin surface layer of the semiconductor. Since the volume of this layer is small, the concentration of free carriers (both electrons and holes) in it becomes very high, which causes a sharp increase in the recombination rate and a decrease in their lifetime. The increase in the recombination rate is also facilitated by the presence of the large number of various impurities and defects in the surface layer, which sharply decrease the mobility of carriers. Thus, intense recombination and strong carrier scattering by the surface defects in the long run nullify the contribution of photo-conductivity to the total conductivity of the sample. The better the surface finishing of the sample, the shorter the waves to which a semiconductor sample is photosensitive.

3.3.6. Impurity Photo-conductivity. Maximum 2 in Fig. 40.a corresponds to the peak of impurity photo-conductivity and is observed in

n—type semiconductors at photon energy sufficient for exciting electrons from the donor level. In a p—type semiconductor this maximum corresponds to the photon energy sufficient for the activation of holes (after receiving this energy, the electrons move from the valence band to the acceptor levels leaving holes in the valence band). It is interesting to note that on both sides of the maximum, the impurity photo-conductivity is zero. While the absence of photo-conductivity in the long-wave region indicates that the energy of incident quanta is too low to ionize the impurity cent res, the absence of photo-current in the region of shorter waves means that the interaction between photons and impurity atoms is resonant in nature: the interaction does not occur when the energy of excitation quanta considerably differs from the energy of activation of impurity centers.

3.3.7. Absorption Spectrum. Fig. 40.b shows the spectral distribution curve for light absorption by a semiconductor (the absorption coefficient K is plotted along the ordinate axis). The absorption spectrum is similar to the photo-conductivity spectrum in many respects, but at the same time it has its peculiarities. For example, in the intrinsic absorption band (region 1′) which, like the band of intrinsic photo-conductivity, is due to band-to-band electron transitions and is bounded by the photo-electric threshold λ_{thr}, the absorption coefficient increases and not decreases with decreasing wavelength. This is because the reasons behind the drop in photo- conductivity in the shorter wave range are not important for light absorption. Therefore, an increase in the energy of excitation photons is accompanied by the transfer of electrons from deeper and deeper energy levels of the valence band to the conduction band.

The impurity absorption band (region with a maximum 2′) is practically identical to the impurity photo-conductivity band (region 2) and is due to absorption of photons generating free impurity carriers during the ionization of the donor or acceptor centers.

Along with the photo-induced processes of light absorption accompanied by the generation of excess free charge carriers, the absorption spectrum may contain regions in which the absorption of light quanta is not associated with a generation of photo-carriers. Region 3 corresponds to the exciton absorption and is the most interesting from the physical point of view.

3.3.8. Exciton Absorption. The concept of excitons was introduced in 1931 by the Soviet physicist Ya. I. Frenkel to explain why the light absorption clearly connected with the excitation of electrons is not accompanied by a change in photo-conductivity. Frenkel's ideas were later brilliantly confirmed experimentally. It turned out that intrinsic atoms of a semiconductor can absorb energy not only when $h\nu \gtrsim W_g$ but also when $h\nu < W_g$. Since in the latter case the energy of the quantum being absorbed is insufficient for the complete ionization of the atom and for the transition of one of the valence electrons to the conduction band, the atom goes into an excited state with one of the valence electrons in the excitation shell. This excited state of the atom was called exciton. Since all the atoms in the crystal lattice of a semiconductor are equivalent, the excited state can be transferred from atom to atom. In this case, the excited atom can return to its ground, unexcited state with the liberated excitation energy being transferred to a neighboring atom in which an electron goes to the corresponding excitation. This process can be treated as the motion of an exciton through the crystal.

An exciton can also be represented as an electron-hole pair, bound by the Coulomb attraction. Clearly, the motion of such a coupled pair reflects only the propagation of the excited state through the crystal and does not lead to a directional transfer of a charge. For this reason, the exciton absorption, having its own band in the absorption spectrum (maximum 3 in Fig. 40.b), does not affect the photo-conductivity (there is no corresponding maximum on the photo-conductivity spectral distribution curve in Fig. 40.a)

Generally, exciton absorption may indirectly influence the photo-conductivity of a crystal. As it moves through the crystal, the exciton may be ruptured either by thermal lattice vibrations or by an additional quantum of light with an energy lower than W_g (in the absence of exciton absorption, this quantum simply could not be absorbed by an intrinsic semiconductor). As a result of this rupture of the exciton, two charge carriers (electron and hole) appear, which make their contribution to conductivity. An exciton can also meet an impurity atom and impart energy to it by detaching an electron and creating an additional charge carrier. On the other hand, the energy of exiton recombination may also cause the transition of an electron from the valence band to the acceptor level. In this case, a free hole appears that is capable of carrying a charge. An exciton may perish when it meets some inactive impurity centre or a lattice defect. The energy of exciton recombination is then transferred to thermal lattice vibrations.

3.4. Luminescence

This name applies to all kinds of "cold" glow. There are about a dozen different types of luminescence. Although they also include thermo-luminescence, which is a glow of some solids (for example, marble or diamond) when heated slightly, this heating is so insignificant that the phenomenon has nothing in common with the rise in body temperature required for visible thermal radiation.

Some crystalline solid bodies begin to glow when rubbed, deformed or fractured. This phenomenon is called triboluminescence. Sometimes luminescence accompanies various chemical reactions in which case it is called chemiluminescence. Another type of luminescence-cathodoluminescence-is observed when some materials are bombarded by electrons. Electric current passing through rarefied gases causes a bright glow whose colour depends on the nature of the gas, and this is called electroluminescence. There are also roentgenolurninescence, radioluminescence, and others. But the most widespread is photoluminescence viz. the luminescence produced by the action of the radiant energy-light-on a material.

3.4.1. Photoluminescence. This phenomenon was discovered about 500 years ago by the Italian shoemaker called Vincencio Casciorola who spent his spare time doing alchemy. But only recently, after the quantum theory of light and the band theory were developed, could photoluminescence be explained sufficiently well. There are two types of photoluminescence: fluorescence and phosphorescence.

The term fluorescence takes its origin from the mineral fluorite (feldspar). The phenomenon consists in that certain materials become sources of radiation under the action of incident light. This radiation can be observed simultaneously with the external irradiation and ceases as soon as irradiation is discontinued. Besides fluorite, this phenomenon can be observed in zinc blende, some glasses and solutions of many chemical compounds.

Unlike fluorescence, phosphorescence is characterized by a certain duration of the luminescence in a material after the source of the radiation is switched off. The intrinsic glow of the phosphorescent substances gradually attenuates with time. Some materials retain their phosphorescence for several months, especially when heated. Phosphorescence is typical of many semiconductors doped with special impurities called activators. Irrespective of the duration of the afterglow, all materials exhibiting luminescence under the action of light are called luminophors. It should be noted that luminophors may glow under the effect of invisible radiation, especially ultra- violet radiation.

Photoluminescence is due to the absorption of the energy of the light and the transfer of atoms from the ground to an excited state. An atom can exist in an excited state for a very short time and must soon return to the ground state, liberating the energy it previously absorbed. However,

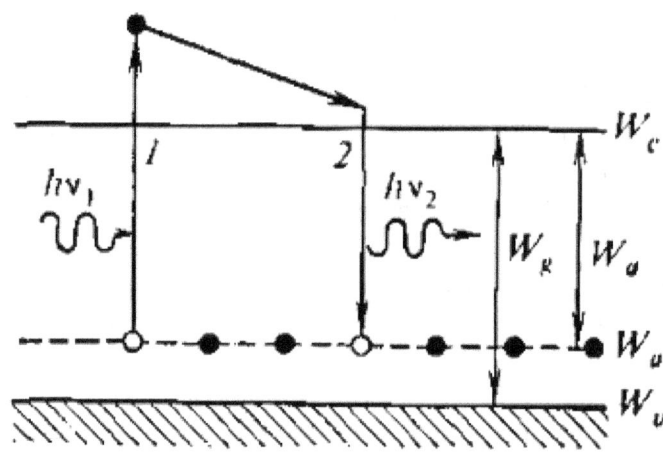

FIGURE 11

although practically all bodies absorb light, only some exhibit luminescence.

3.4.2. Fluorescence. The simplest diagram of a fluorescent luminophor is shown in Fig. 41. A peculiar feature of this diagram is the presence of the activator level W_a. This level appears when a crystal is doped with special impurities, viz. activators. Activator impurities differ in their properties from donor and acceptor impurities. The energy levels associated with activator impurities lie near the valence band, i.e. in the same place as the energy levels of acceptor impurity atoms, which trap electrons from the valence band. However, the activator atoms differ from acceptor atoms in that they do not trap electrons.

On the contrary, being in all unexcited state, they rather exhibit the properties of donor atoms and may give their electrons to the valence band. At any rate, the energy required for detaching a valence electron from the activator atom is less than that for an intrinsic semiconductor atom (in Fig. 41, this property is illustrated in that the separation W_a between the activator level and the conduction band is smaller than the forbidden gap W_g). Thus, while creating energy levels in the region typical of acceptors, the activator impurity has some of the properties of a donor. Activator levels may also appear during the formation of crystals in natural conditions. For this reason some minerals exhibit photoluminescence.

When a fluorescent crystal is illuminated by light with the energy of photons $hv_1 > W_a$, the photons are absorbed by the activator atoms which become ionized. Electrons going from the W_a level to the conduction band (transition 1 in Fig. 41) become free conduction electrons (photoluminescence is always accompanied by photoconductivity). While travelling in the interstitial space, an electron may meet already ionized activator atom and recombine with it. When the electron returns from the conduction band to the W_a level (transition 2), a photon with the energy $h\nu_2$ is emitted, which is perceived as fluorescence. During its voyage through the interstitial space, the electron may spend a part of its energy by being scattered by the lattice atoms performing thermal vibrations or by various defects. Therefore, the energy of the photon emitted in fluorescence cannot exceed the energy of the absorbed photon. Hence;

$$\nu_2 \leqslant \nu_1, \text{ or } \lambda_2 \geq \lambda_1$$

These relations express the rule established experimentally by the English physicist G. Stokes. According to this rule, the wavelength of light emitted by a fluorescent material cannot be less than the wavelength of the source of excitation. In other words, if a body capable of fluorescence is illuminated, say, by blue light, it will emit green, yellow, or red light.

The time the electron spends in the conduction band, i.e. the time between electron excitation and recombination, turns out to be very short. In pure semiconductors containing no impurity atoms besides activators this time may only amount to 10^{-9} sec. Hence, fluorescence is only observed when the crystal is excited by the primary radiation and ceases almost immediately after the radiation source is switched off.

3.4.3. Phosphorescence. In order to obtain more or less persistent glow of a luminophor after the excitation has been discontinued, a semiconductor crystal. must be doped, in addition to activators, with impurity sites that can trap electrons. The energy levels created by such impurities lie near the bottom of the conduction band (W_{im} in Fig. 42), and in their physical properties these sites are similar to trapping centers and trapping sites (see Sec. 3.1)

The initial stage of luminophor excitation and the creation of photoconductivity in phosphorescent materials are the same as in the case of fluorescence. By absorbing photons with an energy $h\nu_1$, the activator atoms are ionized and give their valence electrons to the conduction band. When a liberated electron meets an ionized activator atom, it may recombine, emitting a photon of energy $h\nu_2$. This is the same mechanism of fluorescence. But electrons can also meet trapping impurity sites (transition 2 in Fig. 42).

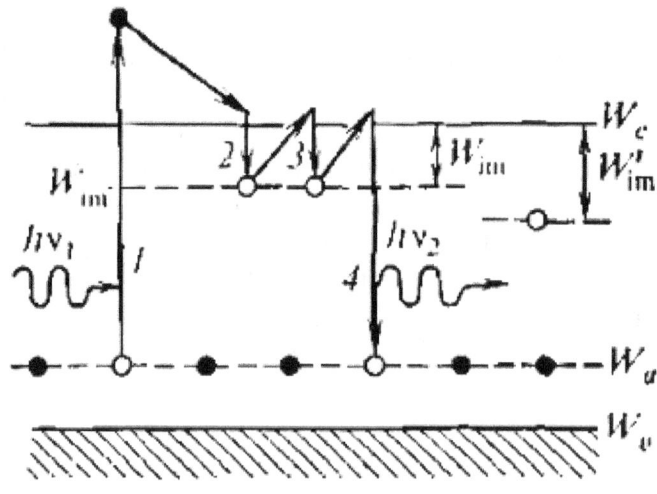

FIGURE 12

In this case, electrons can no longer move through the crystal and thus are excluded from photo-conduction. Moreover, they cannot recombine with the activator ions (a transition from the level W_{im} to the level W_a is not allowed). The electron car return to the conduction band (transition 3) due to thermal lattice vibrations (if the energy of thermal motion of atoms in the crystal is comparable with the energy of activation of the trapping site, i.e. if $W_{im} \simeq kT$). After this, the electron can again be trapped or it can recombine with an ionized atom of the activator (in the latter case, it emits a photon of energy $h\nu_2$ transition 4).

Thus, the fate of a trapped electron does not depend on the presence of the primary excitation radiation. Hence, both residual photoconduction and the ability of a crystal to emit secondary radiation (phosphorescence) can be retained for quite a long time after the source of excitation is switched off. The duration of the afterglow is determined by the depth of the level W_{im} and by the temperature T of the crystal. If the trapping level is near the bottom of the conduction band, and if the activation energy for the trapping site at a temperature T is comparable with the energy of thermal lattice vibrations ($W_{im} \simeq kT$), the afterglow will not last long. At a deeper trapping level (when $W_{im} > kT$), the duration of the afterglow may be considerable, since in order to liberate an electron from the trap in this case the energy of fluctuations of thermal vibrations must be enough for transferring the electron to the conduction band, and this occurs not very often.

Luminescence can be stimulated by heating a crystal (thermal luminescence). This is observed when the energy of the thermal vibrations becomes comparable with W_{im}. By gradual heating a sample, it is possible to determine the depth at which the trapping levels lie in the forbidden band of the sample. Luminescence bursts are observed at temperatures satisfying the condition $W_{im} = kT$. For example, for a crystal whose energy level diagram is shown in Fig. 42, the first afterglow maximum will appear upon heating to a temperature T for which $kT = W_{im}$, and the second maximum, at T' satisfying the condition $kT' = W'_{im}$. Since the W'_{im}, level lies deeper ($W'_{im} > W_{im}$) and requires higher energy for its excitation, $T' > T$. If the trapping level lies far from the bottom of the conduction band, the luminescent properties of the crystal can be retained for an indefinitely long time. Some crystals preserve their luminescence ability upon heating for millions of years (from the time they were formed until nowadays).

CHAPTER 4

Contact Phenomena

4.1. Work Function of Metals

The concentration of free electrons in metals is very high (of the order of $10^{23} cm^{-3}$). These electrons randomly move through a conductor and come continuously to its surface. If, however, the metal is not heated, electrons practically do not leave the sample. This can be explained as follows.

Every free electron moving in the interstitial space of a metal interacts with electrons surrounding it, as well as with positively charged atomic cores which form the crystal lattice. Owing to the uniform distribution of charged particles in the bulk of the crystal, the resultant of all the forces acting on the electron is equal to zero. But when the electron reaches the surface, and the more so when it leaves the crystal, the uniformity of the distribution of the charged particles around it is violated, resulting in forces preventing the electron from leaving the crystal. Two phenomena hindering the electron escaping can be distinguished: the double electric layer formed at the boundary of the metal and the so called electrical image forces.

4.1.1. Double Electric Layer. Free electrons in a metal have rather high kinetic energies even at absolute zero. However, the attraction to the positively charged lattice sites prevents the electron from leaving the metal completely. For this reason, the metal is surrounded by the electron cloud (Fig. 43). The double electrical layer that is thus formed at the surface can be considered as a parallel plate capacitor one of whose plates is formed by the positive surface

ions (their charge being uncompensated after the departure of electrons) and the other plate being formed by the electron cloud in the form of a thin layer. Obviously, the field intensity inside such a capacitor can be assumed constant. If we denote the separation between these plates by a, the force F_1 acting on an electron can be expressed as

$$F_1 = \frac{e^2}{4a^2}$$

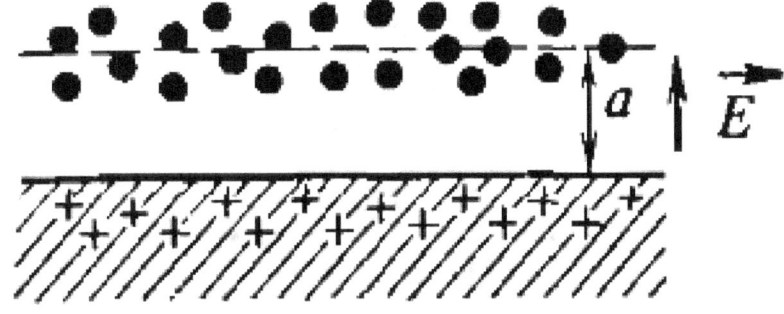

FIGURE 1

and the energy W_1, required for the electron to pass through the double layer will be given by

$$W_1 = F_1 a = \frac{e^2}{4a}$$

4.1.2. Electrical Image Force. Having passed through the double layer, the electron is still subjected to the influence of the metal. Further movement away is hindered by the electrical image force: an electron at a distance x from the surface is acted upon by a force which can be defined as the force of the interaction between the electron and a particle having a positive charge equal in magnitude to the electron's charge and lying within the metal at the distance x from the surface (Fig. 44). According to Coulomb's law, the electrical image force acting on the electron in a vacuum can be expressed by the formula

$$F_2 = \frac{e^2}{4x^2}$$

If we assume that the force F_1 continuously changes into the force F_2 as the electron leaves the double layer, we can represent the force acting on the electron as it moves away from the surface of the metal by the curve shown in Fig. 45. The energy required to overcome the image force is

$$W_2 = \frac{e^2}{4a}$$

4.1.3. Total Work Function. An electron leaving a metal must overcome the potential barrier shown in Fig. 46. Part of this barrier (in the region from 0 to a) is created by the double layer forces, while the other part (in the region from a to ∞) by the image forces. The total height

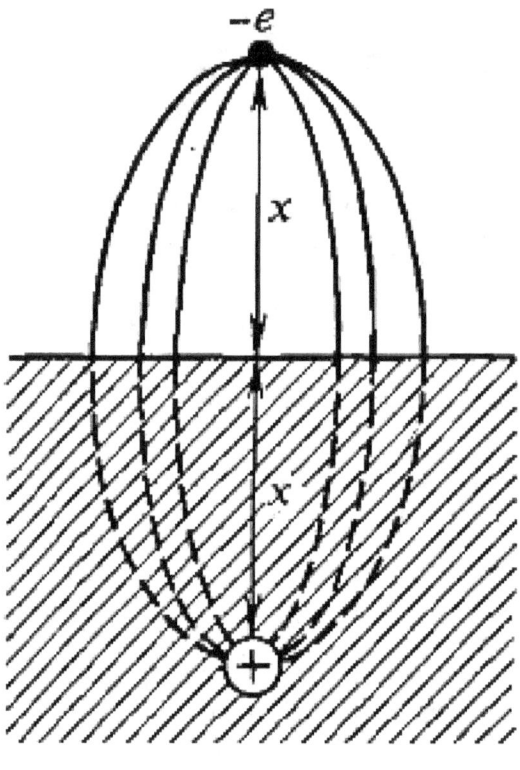

FIGURE 2

W_0 of the potential barrier is equal to the work which must be performed by the electron on its way from the metal's surface to become completely separated from the metal. The quantity W_0 is called the total work function. The experimental values of the work function for different metals lie in the range from 3 to 20 eV.

The shape of the potential barrier can be treated as a curve describing the variation of the potential energy of the electron as it moves away from the metal surface. Indeed, the electron inside a metal has a kinetic energy W_k (consider, for example the electron denoted by 1) and can move in any direction without a change in energy. If, however, it crosses the interface between the metal and a vacuum, its kinetic energy, in accordance with the law of the conservation of energy, will be transformed into potential energy. For example, if the electron is a distance x_1 away from the interface, part of its kinetic energy W'_k will be transformed into potential energy W'_p. Since the electron at the point x_1 still has a part of its kinetic energy W'_k, it can move further away from the surface. However, at a distance x_D its entire kinetic energy is transformed

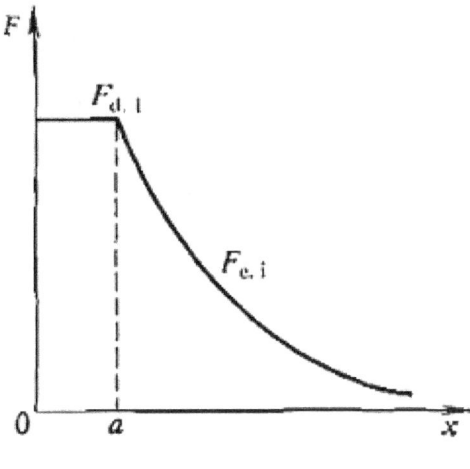

FIGURE 3

into potential energy (point D on the potential energy curve). At this point, the electron's velocity is zero, and it returns to the metal.

In order to be able to leave the metal completely, an electron must have a kinetic energy not lower than the height W_0 of the potential barrier. For example, the electron 2 in Fig. 46 can do this. Moreover, since the initial kinetic energy of this electron exceeds the total work function, even after leaving the metal it will retain some velocity corresponding to the remaining excess kinetic energy:

$$W_{k,exe} = \frac{mv^2}{2}$$

Usually, the energy level corresponding to a stationary electron in vacuum far enough away from the metal surface not to be affected by it is taken as the zero energy level rather than the energy level of a stationary electron in the bulk of the metal. In Fig. 46, the reference level which is approached asymptotically by the electron potential energy curve is denoted as BB. In this case, the bulk of metal is a potential well of depth W_0 This choice of the zero level means that electrons inside the metal and below the BB level have negative energy. Then the total work function is defined as the work done by the electron initially lying on the bottom of the potential well in order to escape from it.

4.2. The Fermi Level in Metals and the Fermi-Dirac Distribution Function

4.2.1. Fermi Level. In spite of the tremendous number of free electrons in a metal, they are arranged in the energy levels of the potential well in a definite order. Each electron occupies a vacancy in the lowest

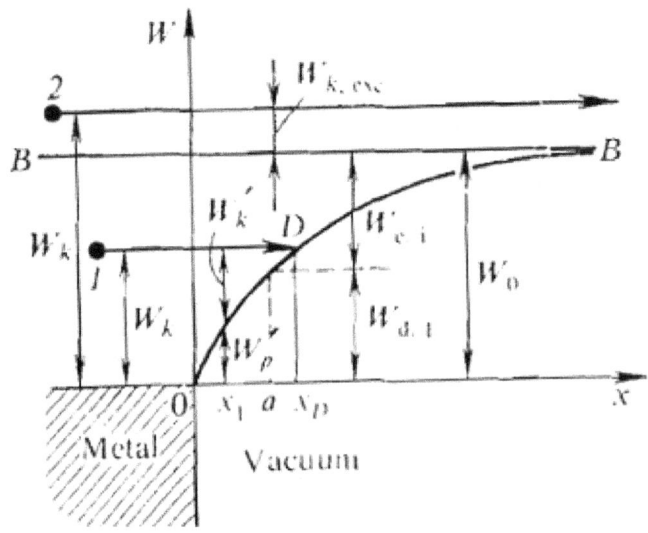

FIGURE 4

possible level. This is quite natural, since every system left alone, i.e. not under any external influence, tends to go to the state with the lowest energy. The distribution of electrons over energy levels obeys Pauli's exclusion principle, which states that no two electrons can exist in completely identical states. For this reason, not more than two electrons having opposite spins can occupy each energy level. As the lower energy levels become filled, higher and higher energy levels are populated. If a metal sample under consideration contains N free electrons, then in the absence of thermal excitation, i.e. at absolute zero ($T = 0$), all the free electrons are arranged pairwise in the $N/2$ lower levels (Fig. 47). The uppermost energy level of the potential well in the metal, occupied by electrons at $T = 0$, is called the Fermi Level and is denoted by μ or W_F. The energy of an electron occupying this level is called the Fermi energy. At $T = 0$, all the energy levels lying above the Fermi level are empty.

Obviously, the work required for electrons in the Fermi level to leave the metal is given by

$$A = W_0 - \mu$$

The quantity A equal to the energy difference between the level BB of extracted electron and the Fermi level is called the thermodynamic work function or simply the work function. It is this quantity that determines how different metals behave when in contact or when a metal-semiconductor junction is formed.

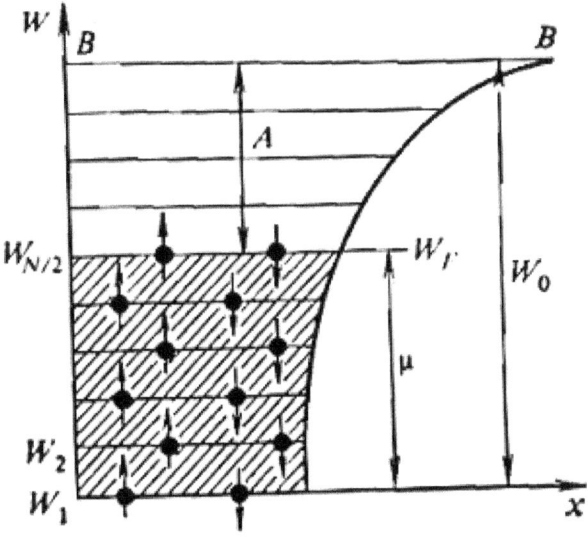

FIGURE 5

4.2.2. Fermi-Dirac Distribution Function. The distribution of particles among various energy levels or states under certain conditions is determined by a distribution function. In the general case, the distribution function describes the probability that a certain level is occupied by particles. If we know for certain that a given level is populated by a particle, it is said that the probability of finding the particle in this level is equal to unity. If, on the contrary, we can state for sure that there are no particles in the level, the probability of finding a particle in this state is zero. However, in many cases we cannot say for certain whether the level is occupied or empty. In this case, the probability that a particle is in this level is greater than zero but less than unity. The higher the probability of finding a particle in the level we are considering, the closer the value of the distribution function for the corresponding state to unity.

If we plot along the x-axis the values of energy corresponding to different levels (from the bottom to the top of the potential well) and probabilities of filling the corresponding levels by electrons along the y-axis, we shall obtain a graph for the Fermi-Dirac distribution function f_{F-D}

At $T = 0$, this graph has the form shown in Fig. 48, which is often called the Fermi step. It can be seen that at $T = 0$ all the levels, including the Fermi level, are occupied by electrons. At the point $W = \mu$, the distribution function drops abruptly to zero. This means that all the levels above the Fermi level are empty at this temperature.

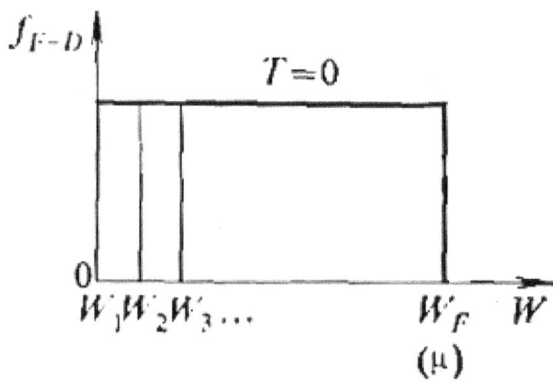

FIGURE 6

4.2.3. Effect of Temperature. At temperatures above absolute zero, the form of the $f_{F-D}(W)$ dependence differs from that shown in Fig. 48. An increase in temperature leads to the thermal excitation of electrons by thermal vibrations of crystal lattice. Due to this excitation, some electrons go from the upper-lying filled levels to empty levels above the Fermi level (Fig. 49). The probability of detecting electrons in these levels now becomes greater than zero. At the same time, the probability that some of the energy levels lying below the Fermi level are full becomes less than

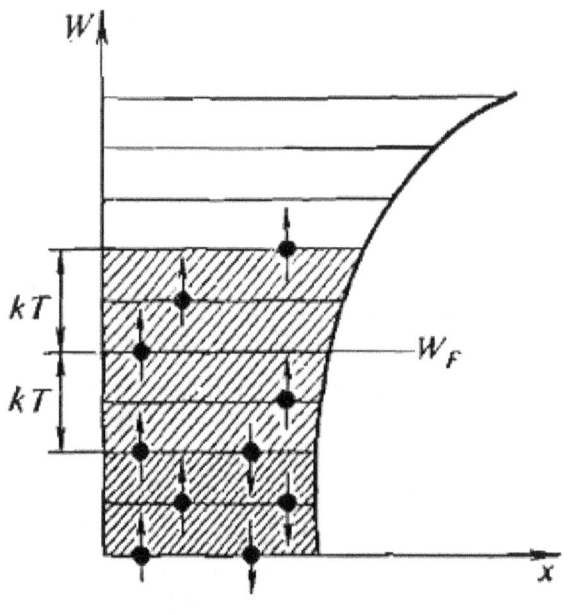

FIGURE 7

unity due to this departure of electrons. Hence, an increase in temperature leads to "blurring" of the boundaries of the Fermi step: instead of the abrupt change from 1 to 0, the distribution function varies smoothly. The dashed lines in Fig. 50 show the distribution function of electrons among levels at $T = 0$, while solid lines correspond to the electron distributions at temperatures above absolute zero. The area of curvilinear triangle under the distribution curve to the right of W_F (area 2) is proportional to the number of electrons that have passed into excitation levels, while the area of the triangle to the left of the value W_F above the distribution curve (area 1) is proportional to the number of electrons leaving the previously filled levels, i.e. the number of vacancies below the Fermi levels. Clearly, the areas of these two triangles are equal, since they correspond to the same number of electrons but considered from different points of view.

It should be noted that at normal temperatures the distribution curve for electrons in a metal is not very blurred. This is because only those electrons lying in the energy levels adjacent to the Fermi level are thermally excited. The depth of the energy levels that are excited can be estimated qualitatively. It is known from molecular physics that kinetic energy of particles due to thermal motion is given by

$$\frac{mv^2}{2} = \frac{3}{2}kT$$

Consequently, the energy transferred to electrons from the lattice atoms thermally vibrating is equal in order of magnitude to kT. At room temperature, $kT \simeq 0.025 eV$, while the Fermi energy for metals at this temperature ranges between 3 and $10\,eV$. This means that not more than 1% of all the free electrons may participate in transitions to higher energy levels in normal conditions. These are precisely the electrons whose energy is close to the Fermi level. As to the electrons populating the energy levels which lie in the bottom of the potential well and separated from the Fermi level by more than kT, they do not take part in thermal excitation, and so their distribution remains the same as it was at absolute zero.

4.2.4. Physical Meaning of the Fermi Level. When we considered the electrical conductivity of solids (see Sec. 1.6), we concluded that this property is related to the possibility of transferring electrons to higher lying energy levels, i.e. it is determined by the ability of electrons to acquire acceleration in an external electric field. In metals at $T > 0$, this is possible only for electrons corresponding to the region where the

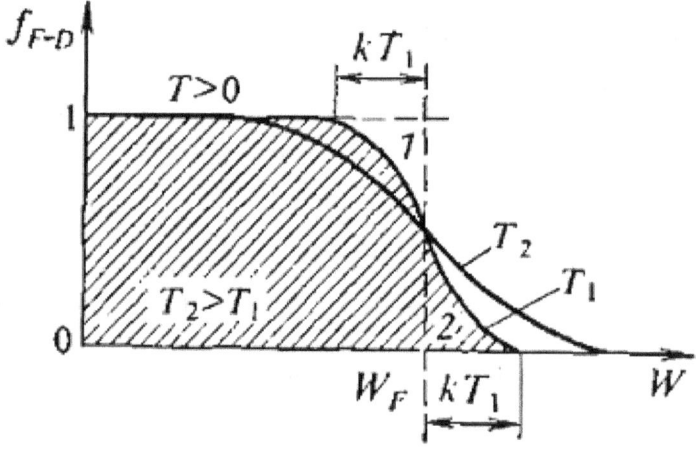

FIGURE 8

distribution function is blurred because the electric fields attainable cannot pull out electrons from the bottom of the potential well and transfer them to unfilled energy levels whose energy is higher than W_F. On the other hand, the deeper electrons cannot go into neighboring higher lying levels because all these levels are filled. Consequently, at $T > 0$, the physical meaning of the Fermi energy is that it is the most probable, or average energy of electrons in metal which can take part in conduction at a given temperature. These electrons are not only responsible for creating conductivity. They also determine the contribution from the electron heat capacity to the total heat capacity of the crystal and to a large extent determine the heat conduction of the crystal.

The Fermi level in metals practically does not change its position with increasing temperature. As the temperature rises, the degree of excitation of electrons increases, and they go to higher levels. At the same time, deeper levels having lower energies become excited. At $T_2 > T_1$, the distribution curve (Fig. 50) is blurred stronger than it is at T_1 and is blurred equally to the left and to the right. Hence, the average energy of electrons taking part in conduction is practically unchanged. It is the more so, since there is a constant electron exchange between the excited levels.

4.3. The Fermi Level in Semiconductors

The subject of the Fermi level in semiconductors is somewhat more complicated than it is in the case of metals. First of all, this is because there are no electrons in the Fermi levels in semiconductors. However, the physical meaning of the W_F (or μ) level remains the same as in the case of metals it is the level which determines the average energy of electrons (or other charge carriers) taking part in conduction. For this reason, the μ level is called the Fermi level in the case of semiconductors as well.

Let us consider the position of the Fermi level in semiconductors.

4.3.1. Intrinsic Semiconductor. In the theory of semiconductors, just as in metals, the level corresponding to the bottom of the conduction band is taken as the zero energy reference.

Thermal excitation of an intrinsic semiconductor is accompanied by transitions of electrons from the valence band to the conduction band. The electrons entering the conduction band do not stay there forever and they soon return to the valence band. At the same time, new electrons from the valence band will replace them in the conduction band. This exchange results in electrons from the upper levels of the valence band taking part in conduction as well the electrons from the lower levels of the valence band. In the reference frame indicated above, the lower level electrons have zero energy, while the energy of the higher level electrons is equal to $-W_g$ (the minus sign indicates that positive values of energy are layed off upwards from the bottom of the conduction band). Thus, the average energy of electrons taking part in conduction is equal to $-W_g/2$. In other words, the Fermi level of intrinsic semiconductors lies in the middle of the forbidden gap (Fig. 51.a).

4.3.2. Impurity Semiconductors. At temperatures close to absolute zero, thermal excitation in the electronic semiconductor can only transfer electrons occupying the donor impurity level W_d to the conduction band. On the other hand, electrons from the valence band cannot take part in conduction, since in this temperature range, the energy of thermal lattice vibrations is insufficient for transferring these electrons into the conduction band ($W_g \gg W_d$). Considering that electrons which arrive to the conduction band are near its bottom and thus have energies close to zero, we can assume (as was the case with an intrinsic semiconductor) that the average energy of electrons participating in conduction is equal to $-W_d/2$. Thus, the μ level in the donor semiconductor at low temperatures lies in the forbidden band at a distance $W_d/2$ from the bottom of the conduction band (Fig. 51b).

In order to create the hole conductivity in p-type semiconductors, electrons from the valence band must be transferred to the acceptor levels lying at a distance W_a from its top. Reasoning in the same way as we did above, we may conclude that the μ level in the hole semiconductor at low temperatures lies between the top of the valence band and the

W_a level of the acceptor impurity. Since we measure the energy from the bottom of the conduction band, we can write

$$\mu = -W_g + \frac{W_a}{2}$$

(see Fig. 51c).

4.3.3. Effect of Temperature on the Position of the Fermi Level.
The position of the Fermi level in an intrinsic semiconductor does not depend on temperature because the processes that determine the conductivity of intrinsic semiconductors are the same regardless of the temperature. Of course, as the temperature and the energy of thermal excitation increase, electrons will go to higher and higher levels of the conduction band, quite far from its bottom. However, at the same time, thermal excitation will transfer electrons from the deepest levels of the valence band to the conduction band. Consequently, the average energy of the charge carriers participating in conduction remains unchanged, and the Fermi level in the intrinsic semiconductor retains its position in the middle of the forbidden gap, regardless of the temperature.

Quite a different situation takes place in impurity semiconductors. As was shown in Sec. 2.2, with increasing temperature the impurity conductivity is replaced by intrinsic conductivity. This leads to a change in the position of the Fermi level. For the sake of definiteness, let us consider a n-type semiconductor.

In the low temperature region, the conduction in electronic semiconductors is determined by transitions of electrons from the donor levels to the conduction band. That is why the Fermi level lies between the impurity level and the bottom of the conduction band. As we showed above, an increase in the temperature leads to the depletion of the impurities, and at $T > T_s$, the donor level is found to be practically empty. But at these temperatures the transitions from the donor level to the conduction band are replaced to an increasingly larger extent by transitions from the valence band. In other words, intrinsic conductivity becomes noticeable. At high temperatures, when $T > T_i$, the conductivity of the semiconductor is almost completely determined by electron transitions from the valence band, and the average electron energy becomes equal to - $W_g/2$ (the μ level in this case is in the middle of the forbidden band). Thus, we see that with increasing temperature the Fermi level in the electronic semiconductor changes its position from - $W_d/2$ to - $W_g/2$ (Fig. 52).

The position of the Fermi level in the hole semiconductor changes with increasing temperature in a similar way. While at $T = 0$ it lies

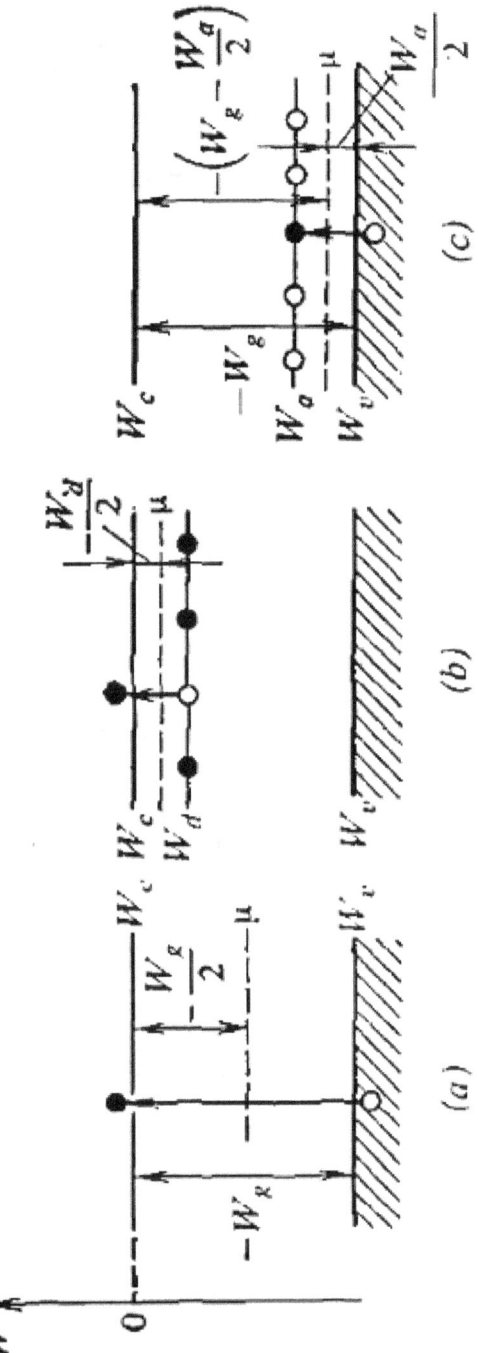

FIGURE 9

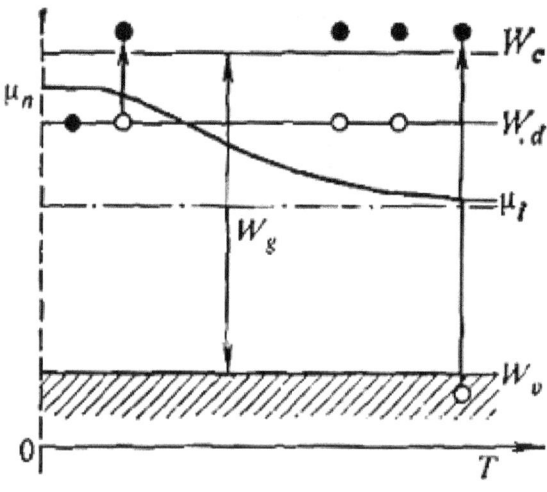

FIGURE 10

between the acceptor level W_a and the top of the valence band, it goes to the middle of the forbidden band as the temperature rises.

4.3.4. The Fermi Level in Degenerate Semiconductors. Unlike common (moderately doped) semiconductors whose Fermi level lies in the forbidden gap, degenerate semiconductors with high doping levels are characterized by μ levels lying in one of the allowed bands: either in the valence band or in the conduction band. By using an *n*-type degenerate semiconductor as an example, we will show how its Fermi level shifts from the forbidden to the conduction band as the concentration of the doping impurity increases.

Till the impurity concentration is not very high and the donor impurity level is narrow and close to the conduction band, the Fermi level lies

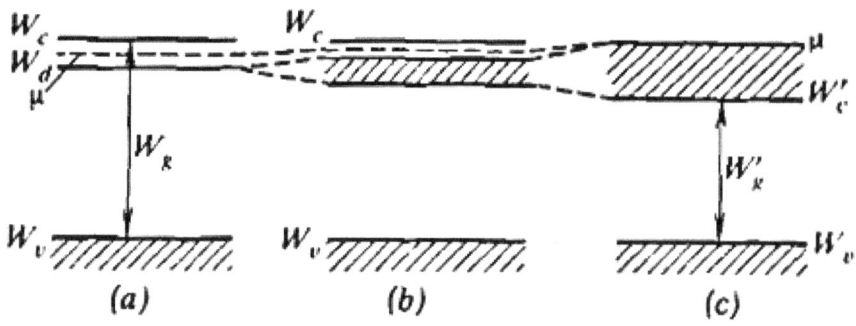

FIGURE 11

in the forbidden gap at a distance - $W_d/2$ from the bottom of the conduction band (Fig. 53a). But when the impurity concentration increases so that the impurity level is blurred into the impurity band, the activation energy of electrons in the donor levels becomes lower because the top of the impurity band is closer to the conduction band than the impurity level in the initial state. Hence, the average energy of the conduction electrons becomes lower, and consequently the Fermi level becomes closer to the bottom of the conduction band (Fig. 53b). At a very high concentration of impurity atoms, the donor band is blurred so much that it joins the conduction band (Fig. 53c). Electrons from the upper levels of the donor band may then freely move over empty levels of the conduction band, and the semiconductor acquires the properties of a metal (there is no need for thermal excitation).

After the bands are joined (at a high concentration of impurity) the Fermi level which at a low doping level was between the top of the impurity band and the bottom of the conduction band, merges with the upper level of the impurity band. Thus, in a degenerate semiconductor the Fermi level plays the same role as it does in metals: it is the uppermost level occupied by electrons at $T = 0$. Since the lower level of the impurity band plays the role of the bottom of the conduction band (level W_c' in Fig. 53c) after the bands have joined, the Fermi level in a degenerate n–type semiconductor turns out to be in the conduction band.

A similar situation occurs when the concentration of acceptor impurity in the hole semiconductor increases. In this case, the acceptor level spreads into the impurity band and joins the top of the valence band. Now the Fermi level of the p–type degenerate semiconductor is found to lie within the valence band, and the higher the doping level, the deeper it lies in the valence band.

4.3.5. The Fermi Level and Work Function in Semiconductors.

The position of the Fermi level determines not only the type of conduction but also properties of the doping material: the lower the activation energy of the impurity sites, the closer the Fermi level is to the appropriate band. Moreover, the position of the Fermi level characterizes the concentration of impurity sites and the temperature of the semiconductor.

Thus, we can say that the position of the Fermi level determines the electrical properties of a semiconductor.

The position of the Fermi level is also related to the work function. At first sight, this relation may seem strange. How can the level in which there are no electrons in the general case affect the magnitude of the

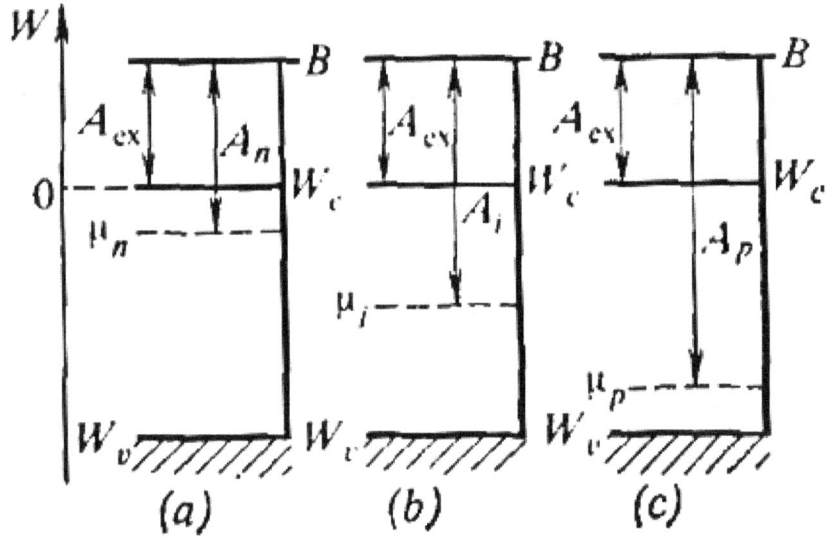

FIGURE 12

work function? Nevertheless a relation does exist and can be explained as follows.

In order to leave a semiconductor, a free electron at the bottom of the conduction band must perform work equal to the energy difference between the bottom and the top of the conduction band, i.e. the level corresponding to the energy of a free electron removed from the semiconductor infinitely. This work A_{ex} is called the external work function (Fig. 54). However, there are no electrons in the conduction band of a non-excited semiconductor. In order to generate them, energy must be spent to transfer electrons to the conduction band from the donor level or from the valence band. In other words, energy is required for generating conduction electrons. Since the average energy of conduction electrons is equal to the Fermi energy, i.e. the energy difference between the Fermi level and the bottom of the conduction band, the work done for creating a conduction electron and withdrawing it from the bottom of the conduction band to beyond the limits of the semiconductor turns out to be equal to the separation between the Fermi level and the top of the conduction band. As in the case of metals, this work is called the thermodynamic work function, or simply the work function.

Figure 54 gives the values of the work function for an n-type (a), intrinsic (b) and p-type (c) semiconductors, respectively. The value of the thermodynamic work function determines the behavior of a semiconductor in all contact phenomena and in all processes associated with the emission of electrons from the crystals.

The external work function A_{ex} is determined by the nature of a crystal and does not depend on the type of impurity introduced. Therefore, the difference between the values of the work function of similar semiconductors containing dis- similar impurities (for example, in the interface between silicon samples with electronic and hole conductivity) can be found from the distances between the Fermi levels and the bottom of the conduction band, without taking into account the position of the top of the conduction band and the magnitude of the external work function A_{ex}.

4.4. The Contact Potential Difference

Let us consider processes which take place when two metals with different work functions A_1 and A_2 are brought close enough so that electron exchange between them becomes possible. This situation can be observed even in the absence of direct contact between metals, since there are always some electrons in metals which can escape from metal sample. It was shown above that at low temperatures, most of these electrons return to the sample. However, if the metals are so close that the surface forces capture "foreign" electrons which are far from "their own" crystal, an intense electron exchange becomes possible between the bodies.

For some time after the metals are placed in contact, this exchange will not be equivalent at all. At the same temperature, the electron flux from the metal with the smaller work function A_1 will be more intense than the opposite flux from the metal with the larger work function A_2. Hence, the surface of the first metal will, be charged negatively, and the surface of the second metal will be positively charged. The potential difference which thus appears will prevent further predominant flow of electrons. When the potential difference U_c between the bodies attains the value at which $|e| U_c = A_2 - A_1$, the number of electrons crossing the vacuum gap between the metals in both directions become equal, and the metal-vacuum-metal system under consideration will be in the state of dynamic equilibrium. The potential difference U_c between metals, corresponding to the establishment of such an equilibrium, is called the contact potential difference.

Let us consider the formation of the contact potential difference from the point of view of the band theory.

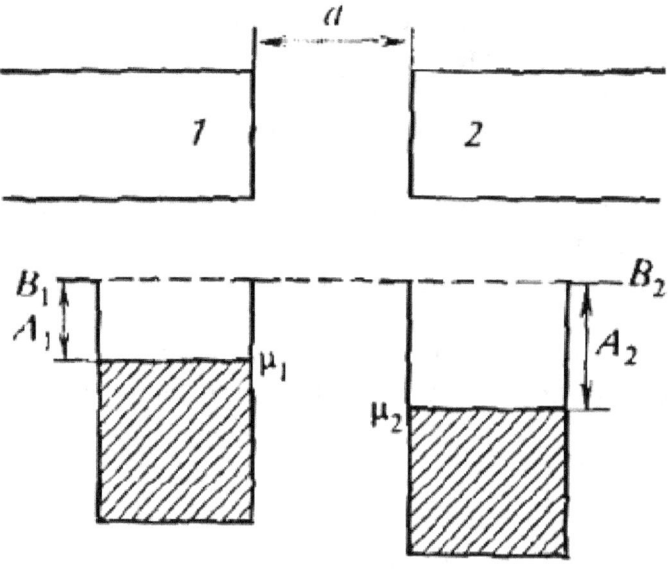

FIGURE 13

Figure 55 shows the Fermi levels and the work functions for the two metals under consideration. The predominant transition of electrons from metal 1 to metal 2 is accompanied by a change in their potentials and the corresponding displacement of their energy levels (on an external scale).

The energy levels of metal 1, which is being charged positively become lower, while the energy levels of metal 2, which is getting positively charged, become higher. The position of the energy levels will change until conditions of electron transfer from one metal to the other become the same. For the electrons occupying the Fermi level, the equilibrium condition is expressed as follows:

$$A_2 = A_1 + eU_c$$

where eU_c is the work which must be done by an electron going from metal 1 to metal 2 in order to overcome the contact potential difference U_c.

Figure 56 illustrates the state of equilibrium between the metals.

There is an energy difference between the Fermi levels of metal 1 and metal 2 before electrons start moving from one metal to the other (Fig. 55), hut once the statistical equilibrium has set in, these levels equalize (on the external energy scale). In this respect, the establishment of statistical equilibrium between the electrons of the two metals is similar to the equalization of the levels of liquid in two connected vessels.

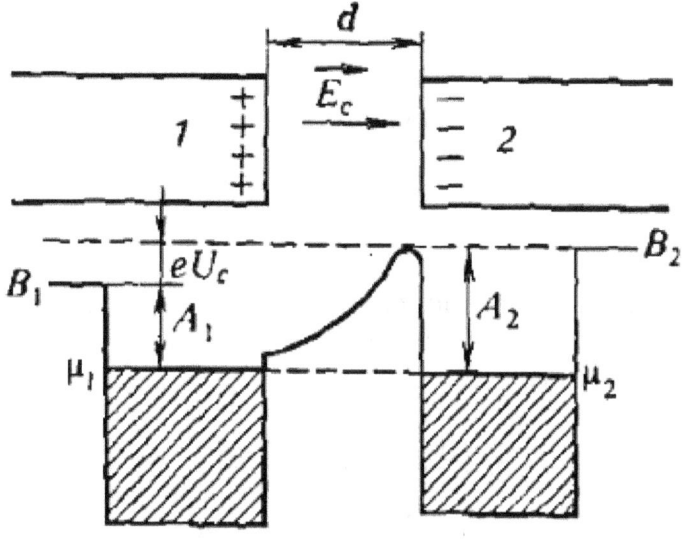

FIGURE 14

If we know the work functions A_1 and A_2 of the two metals, we can define the contact potential difference as follows:

$$U_c = \frac{A_2 - A_1}{e}$$

this value ranges from tenths of a volt to several volts, depending on the choice of the pair of metals.

The equalization of the Fermi levels when two bodies are in contact is a necessary condition for the statistical equilibrium of free charge carriers not only for two metals but also for the contact of any two materials (semiconductors or dielectrics).

The electric field created by the contact potential difference when metals are placed in contact is localized at their interface. In order to verify this, let us assess the number of electrons which pass from one metal to the other as the equilibrium contact potential difference is established. Suppose that $U_c = 1\,V$. Since the minimal distance d to which two crystals can be brought cannot be less than the inter-atomic distance in the crystal lattice, we shall take this distance equal to the lattice constant $a_0 \simeq 0.3\,nm$. The surfaces of two metals in contact can be treated as the plates of a parallel plate capacitor. After the contact potential difference U_c has been established, the electric field intensity in the gap is given by the formula

$$E_C = \frac{U_c}{d}$$

On the other hand, the field intensity is related to the surface charge density σ via the relation

$$E_C = \frac{\sigma}{\varepsilon_0}$$

where ε_0, is the permitivity of a vacuum. Hence, the number n of electrons passing from one metal to the other through the unit interface area can be found thus:

$$n = \frac{\sigma}{e} = \frac{\varepsilon_0 E_c}{e} = \frac{\varepsilon_0 U_c}{ed}$$

Substituting in numerical values, we find that

$$n \simeq 2 \times 10^{13} cm^{-2}$$

On the other hand, the number of free electrons per $1 cm^2$ of the metal surface is of the order of magnitude of $10^{15} cm^{-2}$. Consequently, even when the gap is the minimum possible, only 2 of the free electrons on the surface create the contact potential difference when they go from one surface to the other. For this reason, the contact electric field is localized in the contact gap and does not exist in the bulk of metals.

4.5. Metal-to-Semiconductor Contact

4.5.1. State of Equilibrium. For the sake of definiteness, let us first consider the electronic semiconductor whose work function is less than that of the metal in contact with it. Suppose that initially there is a vacuum gap between the semiconductor and the metal, and the gap is sufficiently narrow for effective electron exchange between the crystals. After the semiconductor and the metal have been placed in contact, the processes on the interface are similar to those when two metals are in contact. The contact potential difference established upon equalization of the Fermi levels is of the order of several volts. The main difference between this case and a metal-to-metal contact is that the depletion of only the external layer of the semiconductor may happen to be insufficient, since in ordinary (non-degenerate) semiconductors the concentration of free carriers is much lower than that in metals.

Let us assess the number of electrons leaving the semiconductor when the separation between surfaces in contact is $1 \mu m$ and in the case of a close contact.

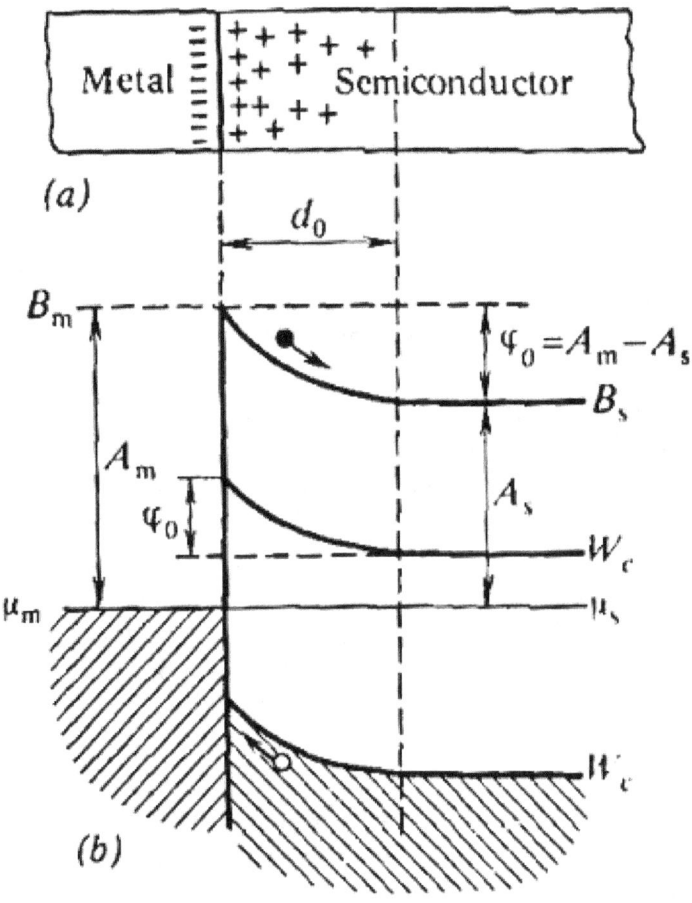

FIGURE 15

By using the formula obtained in the previous section, we find that in order to create the con- tact potential difference of $1\,V$ and $d = 1\,\mu m$, about 5×10^7 electrons must leave each square centimeter of the contact surface of the semiconductor to the metal. In a semiconductor with a moderate doping level, the density of free electrons at room temperature is usually about $10^{15} cm^{-3}$. On the other hand, the number of free electrons in the outer mono-atomic layer is of the order of magnitude of $(10^{15})^{2/3} cm^{-3}$. Hence, less than 1 of electrons contained in the mono-atomic layer of the semiconductor are needed to create the contact potential difference in this case. Clearly, the contact electric field does not embrace either the metal or the semiconductor but is almost completely concentrated in the vacuum gap.

4.5.2. Barrier Layer. A quite different situation occurs when there is close contact between a metal and a semiconductor, the separation between them being of the order of the lattice constant a_0 (for germanium, $a_0 \simeq 0.5nm$). In this case, in order to create the contact potential difference of $1V$, about 10^{13} electrons must pass through each square centimeter of interface. The arrival of this quantity of electrons to the metal will not affect its volume properties much because the total number of transition electrons amounts to only 1% of the number of free electrons contained in the mono-atomic layer of the metal.

All these electrons are accommodated in the surface layer of the metal, close to the contact area. In the semiconductor, the pattern is different. Since there are only 10^{10} free electrons in $1cm^2$ of the mono-atomic" layer of a moderately doped semiconductor, the transition of 10^{13} electrons can only be provided if all the free electrons in 1000 atomic layers of the contact region of the semiconductor will take part in the transition (Fig. 57.a). The region where only fixed uncompensated positive donor ions are left is a typical dielectric. The thickness of this region for most commercially used semiconductors ranges between and $10^{-3}cm$ 10^{-5} and exceeds many times the mean free path of carriers. As this depleted region contains no free carriers and has much greater thickness in comparison with the mean free path, it has large resistance and is thus called the barrier layer.

It is clear that in this case, the electric field is no longer localized in the vacuum gap in the interface, but also exists in the barrier layer, which has a thickness d_0. Moreover, since the width of the vacuum gap is many times less than the thickness of the barrier layer, we can assume that the electric field is only created in the depleted layer of the semiconductor.

4.5.3. Bending of Bands. The penetration of the contact field into a semiconductor (and hence the difference potentials in different cross sections of the semiconductor) leads to the displacement of energy levels and bands towards higher or lower energy values, these displacements being different for different cross sections (Fig. 57.b).

The behavior of free carriers in the electric field existing in the contact layer of a semiconductor can be visualized qualitatively in the following way. Imagine that electrons are heavy balls, and holes are bubbles of gas in a liquid. An electron would then slide down a bent energy level towards its lowest point, while a hole coming to the surface will move along the level it is occupying towards its elevation. According to this model, for the contact under consideration electrons in the conduction band near the interface must slide down to the bottom of the band, thus leaving the contact region depleted of majority carriers. On the other hand, holes in the valence band must rise towards the interface, thus enriching the contact region in minority carriers. Although the arrival of additional holes to the contact region increases the hole concentration, it cannot noticeably compensate for the departure of the majority carriers, since the concentration of the minority carriers is always much lower than that of the majority carriers. For this reason, the barrier region is characterized by a high resistance.

4.5.4. Anti-barrier Layers. A barrier layer can thus he formed in the case of contact between an n—type semiconductor and a metal, but only if the work function of the metal A_m is larger than the work function A_s of the semiconductor ($A_m \gg A_s$,). If, however, $A_m < A_s$, the reverse situation is observed. In this case, after the crystals are placed in contact, the electron flux from the metal will prevail over the opposite flux from the semiconductor until the dynamic equilibrium sets in. The semiconductor will thus acquire a negative charge and the metal, a positive charge. The direction of the electric field in the boundary region of the semiconductor and the corresponding changes in the semiconductor potential will be such that the energy levels and hands in the semiconductor are curved upwards away from the interface with metal. Therefore, electrons arriving from the bulk of the semiconductor "slide" to the interface, thus enriching this region with majority carriers (Fig. 58). The increase in the majority carrier concentration in the contact layer decreases its resistance. For this reason, such a layer is called an anti-barrier layer.

Similar processes occur when a metal and a p—type semiconductor are placed in contact. The only difference is that the barrier layer in this case appears when the work function of the metal is less than that of the semiconductor ($A_m < A_s$). On the other hand, if $A_m > A_s$, an anti-barrier layer is formed.

4.6. Rectifier Properties of the Metal-Semiconductor Junction

The most interesting types of contacts considered above are those associated with the formation of barrier layers, since such contacts have pronounced unipolar conductivity. If a closed circuit contains a junction with a barrier layer, the current intensity at one polarity (the forward direction) is high, while in the case of opposite polarity of the battery, the current intensity is many times lower (cut-off direction). Let us consider this phenomenon by using as an example the contact between an n—type semiconductor and a metal, with which we are familiar. The energy band diagram of such a junction in equilibrium is shown in Fig. 59.a.

First of all, it should be noted that unlike the diagram in Fig. 57, this diagram does not show the energy levels corresponding to the energy

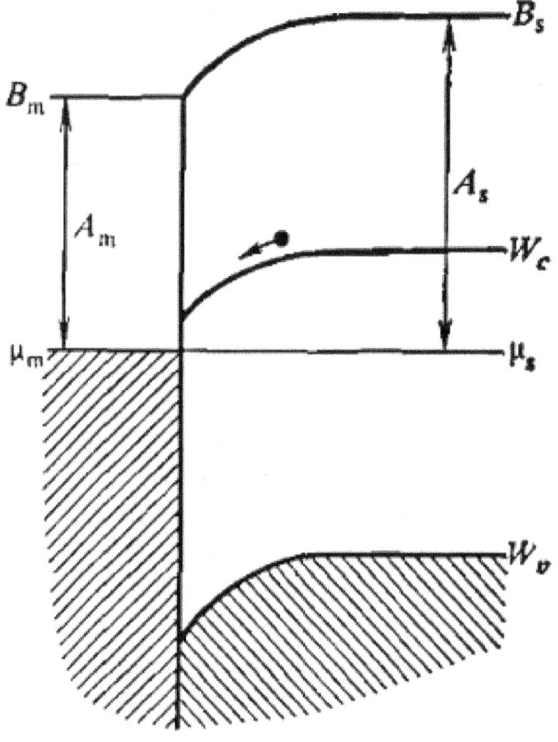

FIGURE 16

of electrons in a vacuum (levels B_m and B_s are absent). Such a simplification can be made when considering contact phenomena because we are not interested in the emission of electrons into a vacuum, and hence there is no need to indicate on the diagram the magnitudes of the thermodynamic work functions A_m and A_s for the metal and semiconductor. On the other hand, the difference in the work functions of the bodies in contact, which determines the conditions under which equilibrium sets in and which characterizes properties of the interface, is shown on the diagram as the height of the potential barrier

$$\varphi_0 = A_m - A_s$$

The potential barrier appearing on the interface regulates the electron flux from semiconductor to metal. When a junction is formed, the number of electrons arriving per second to the metal from the semiconductor greatly exceeds the opposite flux, viz. the number of electrons per second from the metal to the semiconductor. This is because the work function of the semiconductor under consideration is less than the work function of the metal. But as the potential difference, and hence the intensity of the electric field, preventing the electrons from going from the semiconductor to the metal, increases, these opposing fluxes equalize. In this case, there is no current through the junction.

As a result of applying an external potential difference, the equilibrium is disturbed and an electric current through the interface appears. The current intensity depends on the polarity of the power supply and the magnitude of the potential difference applied.

4.6.1. External Potential Difference Applied in the Forward Direction. Let us first consider the case when the external potential difference applied to the metal-to-semiconductor contact is opposite to the contact potential difference. This means that the semiconductor receives a negative potential with respect to the metal. As a result, all the energy levels of the semiconductor, including the Fermi level, are raised by eU (where U is the applied potential difference). Such a displacement of levels (Fig. 59.b) leads to the lowering of the potential barrier which must be surmounted by electrons going from the semiconductor to the metal. Now, this barrier has a height

$$\varphi = \varphi_0 - eU$$

The lowering of the potential barrier φ results in a sharp increase in the number of electrons going from the semiconductor to the metal. On the other hand, the potential barrier for electrons moving from the metal to the semiconductor remains the same as it was in the absence of the external field. Thus, the balance of currents through the junction is disturbed: the electron flux from the semiconductor to the metal is several times more than the opposite flux of electrons from the metal. This disturbance of the balance between the fluxes leads to a current through the junction from the metal to the semiconductor, the intensity of this current being greater the larger the potential difference U applied to the junction.

Since the resistance of the bulk of the semiconductor (and the more so of the metal) is many times lower than the resistance of the barrier layer, the potential drop occurs almost completely across the contact region of the semiconductor. As a result, the potential difference in the contact region and the intensity of the field pushing electrons out of this region into the bulk of the semiconductor decrease. Naturally, the energy bands become less bent and the region depleted in the majority carriers becomes narrower. In other words, the application of the external potential difference in the case under consideration not only lowers the potential barrier φ but also decreases the thickness d_0 of the barrier

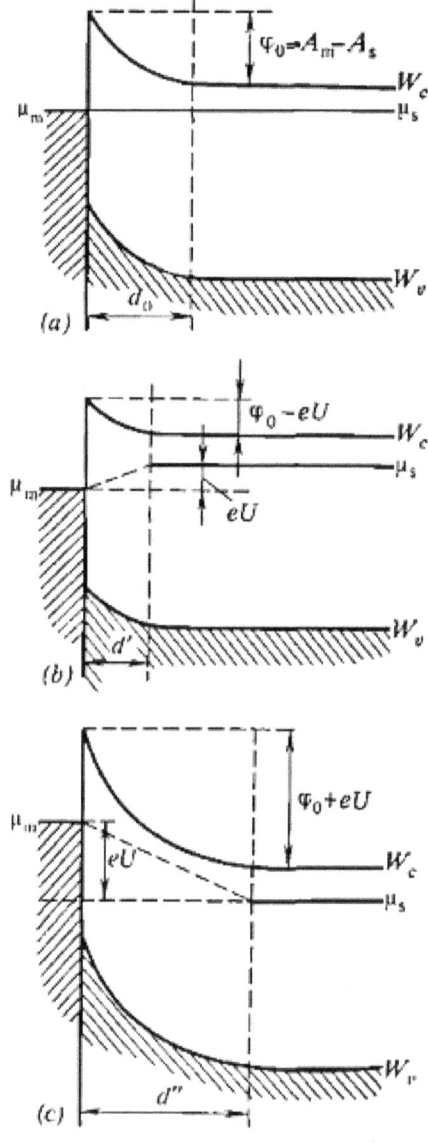

FIGURE 17

layer, thus decreasing the resistance of the junction. These two factors cause an increase in the intensity of current flowing through the junction.

In this case, the external potential difference is said to be applied in the forward, or through direction.

4.6.2. External Potential Difference Acting in the Cut-Off Direction. Quite the opposite situation occurs in the contact under consideration when the external potential difference is applied in the same direction as the contact potential difference. In this case, the external potential difference is said to be applied in the reverse, or cut-off direction.

The application of this potential difference lowers all the energy levels in the semiconductor, including the Fermi level, by eU (Fig. 59.c) with respect to the position they have in equilibrium conditions. In this case, the height of the potential barrier which prevents electrons from going from the semiconductor to the metal increases and becomes equal to

$$\varphi = \varphi_0 + eU$$

while the intensity of the electron flux from the semiconductor to metal decreases. Since the potential barrier for electrons going from the metal to the semiconductor remains unchanged, the balance of electron fluxes through the interface will be disturbed: the electron flux from the metal will be greater than the opposite flux, and hence a current will flow through the junction from the semiconductor to the metal.

It should be noted that the disturbance mechanism for the balance of opposite electron fluxes through the metal-semiconductor junction are different for the external potential differences applied in the forward and reverse directions. In the first case, the balance is disturbed because of an increase in the flux of electrons from the semiconductor to the metal. The current appearing in the circuit is due to the prevalence of this flux over the opposite electron flux which remains the same as it was in the absence of the external potential. This prevalence is the greater the larger the applied potential difference. For this reason, the current may increase practically indefinitely when external potential difference increases within certain limits.

On the other hand, when the external potential difference acts in the cut-off direction, the balance between the electron fluxes is disturbed due to a decrease in the electron flux from the semiconductor to the metal. Consequently, the current appears because the electron flux from the metal to the semiconductor is not compensated by the opposite flux. In this case, the higher the cut-off voltage, the less is the compensation. When the potential barrier for electrons going from the semiconductor to the metal becomes so high that the transfer of electrons to the metal practically ceases, the intensity of the current through the junction reaches its maximum value. No further increase in the cut-off voltage will lead to

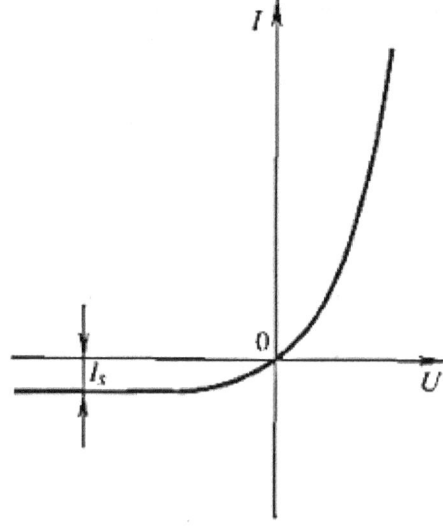

FIGURE 18

an increase in the current intensity. This is because the height of the potential barrier for electrons moving from the metal to the semiconductor does not depend on, the cut-off voltage.

The maximum current flowing through the junction in the cut-off direction is called the saturation current and is denoted by I_s.

The increase in the reverse current is also limited because the external potential difference (distributed mainly in the contact region of the semiconductor) is added to the contact potential difference, and so the electric field pushing the majority carriers (electrons) into the bulk of the semiconductor increases. This leads to an electron depletion of a wider contact region in the semiconductor, and hence to an increase in the barrier layer thickness and its resistance.

4.6.3. Current-Voltage Characteristic of the Metal - Semiconductor Junction. The current-voltage characteristic of a metal-semiconductor junction is shown in Fig. 60 and illustrates the non-symmetric behavior of this junction with respect to the dependence of current on the external voltage. It can be seen from the figure that the junction exhibits a well-defined unipolar conductivity: it unlimitedly conducts current in the forward direction and almost does not in the reverse direction.

4.6.4. Ohmic Contact. Unipolar conductivity is not exhibited by all junctions between metals and semiconductors but only by those in which

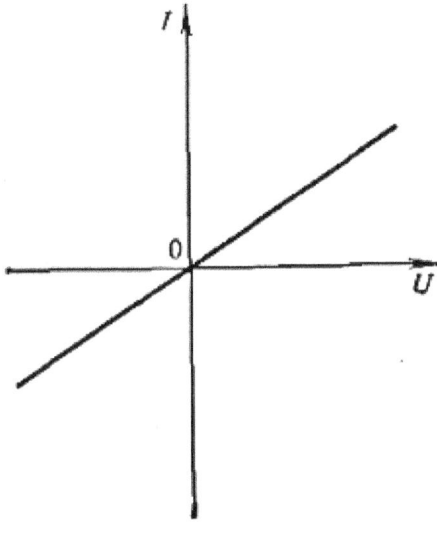

FIGURE 19

a barrier layer is formed. Junctions with anti-barrier layers do not possess this property.

Junction with anti-barrier layers are also widely used in radio-engineering and electronics for connecting various devices. The main requirement to the joining contacts is that they should riot distort the shape and nature of a signal. Their current-voltage characteristic should therefore be linear and contacts with anti-barrier layers satisfy this requirement. Ohm's law is observed for them and so they are called ohmic contacts. The typical current-voltage characteristic of the ohmic contact is shown in Fig. 61.

4.7. $p-n$ Junction

Most semiconductor devices used in modern technology contain the junction between two impurity semiconductors with different types of conduction, which is called $p-n$ junction.

4.7.1. Methods of Obtaining $p-n$ Junctions. A junction between the electronic and the hole semiconductor can be obtained by placing two samples having different types of conductivity into close contact. However, this method cannot be used to create a device in practice because of the presence of various defects and impurities, and first of all oxide films which always cover semiconductor surfaces, sharply change properties of the interface between semiconductors. Hence, in order to obtain a $p-n$ junction with controlled and permanent properties, it is necessary to obtain it in the form of the inner interface on which the

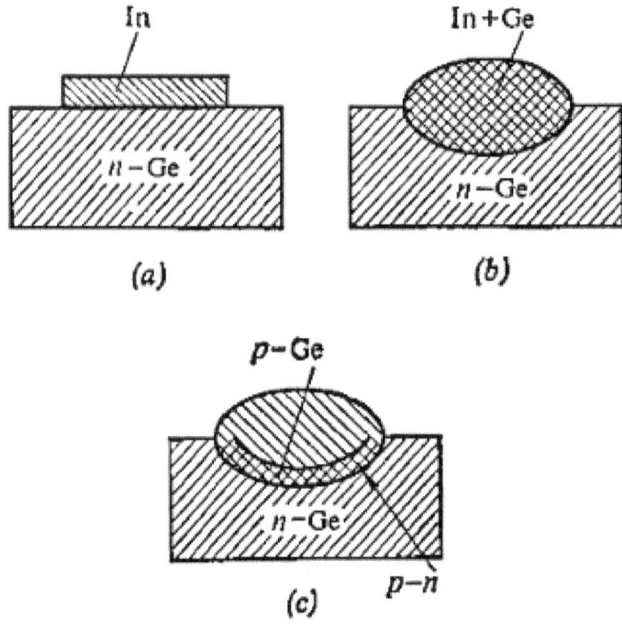

FIGURE 20

semiconductor of one type continuously transforms into the other type of semiconductor. At present, there are many methods of preparing $p-n$ junction. Here, we shall only mention two of them: the method of alloying and the diffusion method.

In the alloying method, a small sample of a trivalent metal (e.g. indium) is put onto a plate of n—type semiconductor (Fig. 62a). This is then placed into a furnace and heated up to $550 - 600\,^\circ C$ in an inert-gas atmosphere. At this temperature, the indium melts and the droplet formed as a result of melting dissolves the germanium (Fig. 62b). After a certain time, the furnace is switched off, and germanium containing indium impurity atoms begins to precipitate from the melt as it cools. If the cooling proceeds sufficiently slow, the precipitating germanium crystallizes in the form of a single crystal whose orientation coincides exactly with that of the substrate single crystal. However, unlike the substrate, which has n—type conductivity,the newly formed germanium region has p—type conductivity. Therefore, $p-n$ junction is formed on the interface between the undissolved part of the substrate and the region that has crystallized (Fig. 62.c). An alloyed $p-n$ junction is characterized by an abrupt change in the type of conduction from the electronic to the hole type, and is thus called an abrupt $p-n$ junction.

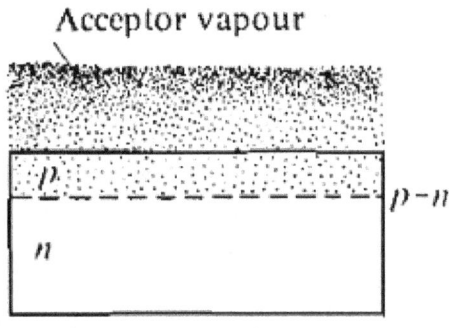

FIGURE 21

In contrast to an alloyed junction. the diffused junction is characterized by a gradual change in conductivity from the $n-$ to $p-$ type, and therefore, it is called the graded $p-n$ junction. This junction is formed by the diffusion of the acceptor impurity from the gaseous or liquid phase into the donor semiconductor, or that of the donor impurity into the acceptor semiconductor.

If the donor semiconductor is used as the substrate, the acceptor atoms penetrating into the bulk of the sample as a result of diffusion first convert it into a compensated semiconductor (i.e. they neutralize the donor impurities and impart to the sample properties of the intrinsic semiconductor), and then as impurity atoms are accumulated turn it into the hole semiconductor. The doping level of the $p-$type semiconductor formed, as well as the depth of penetration of the acceptor atoms into the substrate, are determined by the temperature and duration of diffusion. The interface between the part of the semiconductor transformed into the $p-$region and the $n-$type substrate that remains uncontaminated by the diffusion forms p-n junction (Fig. 63). Naturally, the interface between the acceptor and the donor semiconductors, which are actually different parts of the same sample, cannot be clear-cut because the change in the conduction type occurs gradually in diffused $p-n$ junctions.

There are symmetric and asymmetric $p-n$ junctions. If the doping levels on different sides of the $p-n$ junction are identical or quite close we have a symmetric $p-n$ junction. If, however, the concentration of one of the impurities, say, the donor, is higher than the impurity concentration on the other side of the junction ($N_D \gg N_A$), the $p-n$ junction is asymmetric. Although asymmetric $p-n$ junctions are the ones mostly used in semiconductor technology, for the sake of simplicity, we shall only consider a symmetric $p-n$ junction and assume that $N_D = N_A$.

4.7.2. p−n Junction at Equilibrium. Let us consider an abrupt $p-n$ junction with a symmetric distribution of impurities on both sides of the interface. Suppose that in the electronic semiconductor on the left of the interface (Fig. 64.a) the concentration of a donor impurity abruptly drops from N_D to 0, while in the hole semiconductor on the right the concentration of the acceptor impurity abruptly increases from 0 to N_A. In the n−type semiconductor, the majority carriers are electrons, while in the p−type semiconductor, they are holes. The appearance of a large number of majority carriers in the contact regions is due to thermal excitation and the ionization of the donor and acceptor centres. At a moderately high (e.g. room) temperature, practically all the donor and acceptor impurity atoms are ionized. Therefore, we can assume that the concentration of electrons in the n−type semiconductor far from the interface is equal to the donor impurity concentration ($n_{n_0} = N_D$), while the hole concentration in the p−type semiconductor is equal to the concentration of the acceptor centers ($p_{p_0} == N_A$). The subscript "0" indicates that the concentration corresponds to the equilibrium state.

In each of the regions in contact, there is a certain number of minority carriers in addition to the majority carriers: there are holes in the n−type region and electrons in the p−type region. Their concentration can be found by using the law of mass action (see Sec. 2.6). Suppose that the impurity concentration is $10^{16} cm^{-3}$ in each of the contact regions. The majority carrier concentration will be of the same value:

$$n_{n_0} = p_{p_0} = N_D = N_A = 10^{16} cm^{-3}$$

If we assume that at room temperature the carrier concentration in the intrinsic semiconductor is $n_i = 10^{13} cm^{-3}$, for the minority carrier concentration (e.g. electrons in the p−type semiconductor) we obtain

$$n_{po} = \frac{n_i^2}{p_{po}} = \frac{10^{26}}{10^{16}} + 10^{10} cm^{-3}$$

The number of holes in the electronic semiconductor will be the same. Thus, it can be seen that the concentration of one type of carrier (electrons or holes) changes 10^6 times as we cross the interface. However, the carrier concentration near the interface cannot change as abruptly as does the concentration of the doping impurity atoms, since the impurity atoms are involved into the crystal lattice and are rigidly bound to it, while carriers are free and can move through-out the crystal. The change in the carrier concentration in the contact region is schematically shown in Fig. 64.b.

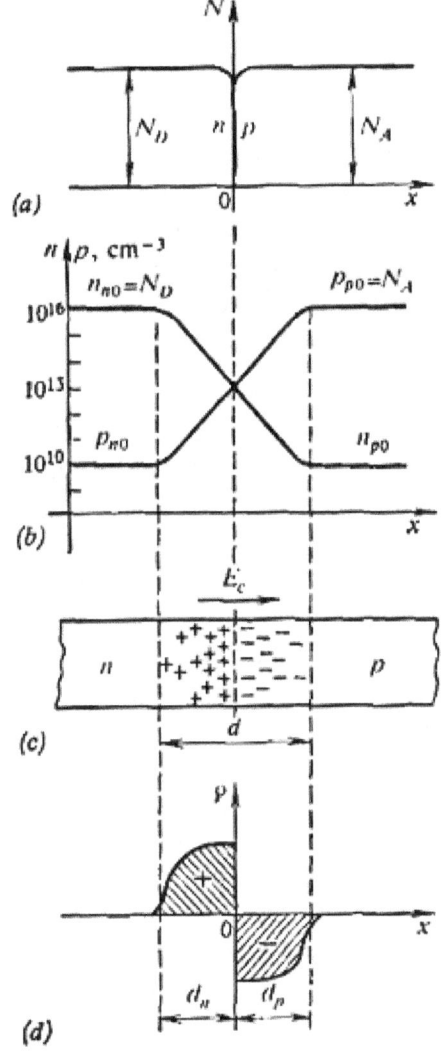

FIGURE 22

However smooth the variation in the concentration may be, the total change turns out to be very large. This leads to intense diffusion of the majority carriers through the $p-n$ junction: electrons go from the n–type to the p–type region and holes diffuse in the opposite direction. As a result of this diffusion, a diffusion current flows through the $p-n$ junction, and the contact potential difference appears.

4.7.3. Diffusion Current. The electrons arriving to the p–type region, which is rich in holes, recombine quite rapidly. Therefore, their concentration in this region practically does not increase and remains at the equilibrium level. At the same time, as a result of continuous thermal generation, new electrons appear in the n—type region in place of those which have left it. Hence, in spite of the diffusion of electrons from the n—type region to the p—type region, the difference in their concentrations and the electron diffusion rate in these regions remain unchanged. This is also true of the diffusion of holes from the p—type to the n—type region. Such a steady-state directional charge transfer through the $p-n$ junction is called the diffusion current. It should be emphasized that the diffusion current created by transition of holes adds up with the diffusion current created by the opposite electron flux.

4.7.4. Contact potential Difference. Conduction Current. The arrival of holes from the p—type semiconductor to the contact region of the electronic semiconductor and, what is more important, the departure of a large number of electrons from this region, which leave uncompensated positively charged immobile ions of the donor impurity, create a positive charge at the interface in the n—type semiconductor (Fig. 64.c). Similarly, the departure of holes from the contact region of the p—type semiconductor, as well as the arrival of electrons from the n—type semiconductor, create a negatively charged region near the interface in the p—type semiconductor. Let us denote by d_n and d_p the widths of these regions in the$n$—and p—type semiconductors respectively. Figure 64.d shows the charge density distribution in the contact regions. The contact potential difference U_c and the electric field which prevents the majority carriers from further diffusion are created between oppositely charged regions. At the same time, this field promotes rather than hinders the transition of the minority carriers through the interface. Indeed, electrons arriving at the $p-n$ junction from the bulk of the p—type semiconductor and the holes from the bulk of the n—type semiconductor are entrained by the contact field and carried to the n—type and the p—type regions, respectively. Thus, in contrast to the diffusion current that appears as a result of majority carriers crossing the interface, reverse current appears, which results from the transition through the same interface of minority carriers in the opposite direction. This current is called the conduction current. Its intensity does not practically depend on the contact potential difference and is solely determined by the thermal generation of minority carriers and the conditions of their transition from the bulk of the semiconductor to the interface.

4.7.5. Band Structure of the $p-n$ Junction at Equilibrium. Figure 65.a shows the .band diagrams of n—type and p—type semiconductors before the carriers start crossing the interface. The electron levels in a vacuum, the two bottoms of the conduction bands and the tops of the valence bands are aligned, while the Fermi level is near the bottom of the conduction band in the n–type region and near the top of the valence band in the p–type region at room temperature. The diffusion flow of the majority carriers, resulting in the stripping of the ionized donor and acceptor atoms and appearance of uncompensated charges in the contact regions, causes the displacement of energy levels. In the n–type region, which receives a positive charge, all the levels are shifted downwards, while in the p–type region, which acquires a negative charge, they are shifted upwards. As was shown above, the levels are displaced until the Fermi levels are aligned (Fig. 65.b). This corresponds to the establishment of the equilibrium state. Owing to unlike volume charges formed in the contact regions on either side of the interface, the potentials of semiconductors change accordingly, and the energy bands in the region of the $p-n$ junctions will be bent. Since the Fermi levels retain their positions with respect to the other energy levels in the bulk of semiconductors sufficiently far from the interface, the bending of the energy bands creates a potential barrier $\varphi_0 = eU_c$. It can be seen from the figure that

$$\varphi_0 = \mu_p - \mu_n$$

or (what is the same)

$$\varphi_0 = A_p - A_n$$

As we showed above, the position of the Fermi level depends on the doping level of a semiconductor: the higher the doping level, the closer the Fermi level is to the appropriate allowed band. For example, the larger amount of the donor impurity introduced into a semiconductor, the closer the Fermi level gets to the bottom of the conduction band. In the limiting case, the Fermi level in the donor region of nondegenerate semiconductor goes right up to the bottom of the conduction band, and in the acceptor region it goes to the top of the valence band. Thus, the maximum height of the potential barrier φ_{0max} at the interface between two non-degenerate dissimilar semiconductor is equal to the forbidden band width W_g. However, in reality the values of φ_0, for the $p-n$ junction between non-degenerate semiconductors are usually less than W_g. For example, for the $p-n$ junction in germanium, for which $N_D = N_A = 10^{16}\, cm^{-3}$, the height of the potential barrier at room temperature amounts to $0.35\, eV$, which is approximately equal to a half the forbidden band width.

The contact electric field controls the transition of carriers through the $p-n$ junction. Not all electrons going from the n–type to the p–type

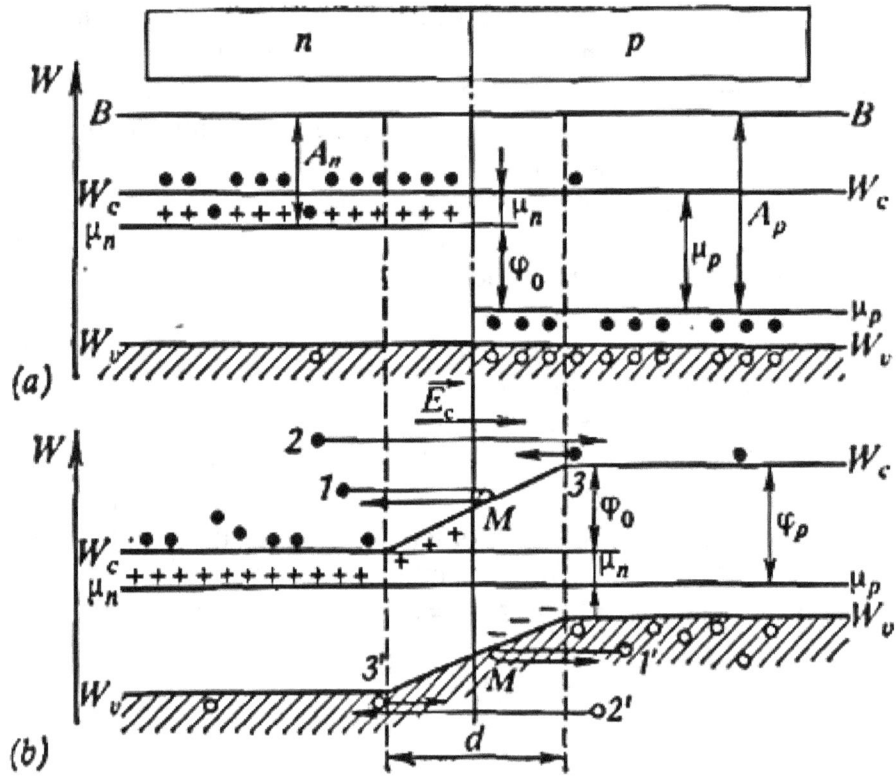

FIGURE 23

region may overcome the potential barrier φ_0,. For example, the electron denoted by 1 in the figure can only surmount part of the potential barrier at the expense of its excess kinetic energy. When it reaches the point M, all its kinetic energy is spent to perform work against the forces of the contact field. The electron stops and is then returned by this field to the bulk of the n–type semiconductor. A similar situation occurs for the hole denoted by $1'$ and arriving from the p–type region. Only those of majority carriers whose energy is higher than φ_0, can surmount the potential barrier. This opportunity is given, for example, to electron 2 and hole $2'$. However, the amount of high-energy electrons and holes is small, because most electrons are near the bottom of the conduction band and the major part of holes, near the top of the valence band. For this reason, the fluxes of majority carriers through the $p-n$ junction are small, although the total number of these carriers is extremely large.

The action of the contact field on the minority carriers is quite opposite. This is illustrated by Fig. 65.b. When electron 3 and hole $3'$ arrive at the $p-n$ junction, they are carried by the contact field to the opposite side of the interface. Their kinetic energies increase at the expense of the work done by the forces of the field. However, despite the favorable conditions for electron transitions from the p—type to the n—type region and for hole transitions from the n—type to the p—type region, the fluxes of the minority carriers are not large, since the total number of minority carriers in each semiconductor is small.

The intensity of the contact electric field almost does not affect the magnitude of the minority carrier fluxes through the interface. Regardless of the intensity of the contact field, any minority carrier reaching the region of the $p-n$ junction will be transferred by this field to the opposite side of the junction. The magnitude of each of the fluxes is determined by the number of minority carriers (electrons or holes) arriving at the junction, i.e. by the number of carriers formed as a result of thermal generation in the immediate proximity of the junction. This is why the intensity of conduction current does not depend on the magnitude of the contact potential difference. The thickness of the semiconductor layer adjoining the $p-n$ junction and participating in the creation of the conduction current is determined by the diffusion length L_n for electrons in the p—type region and L_p for holes in the n—type region, i.e. by the mean distance covered by the electrons and holes during their lifetimes. Electrons appearing due to thermal generation in the bulk of the p—type semiconductor at distances from the $p-n$ junction exceeding L_n (as well as holes appearing in the n—type semiconductor at distances from the $p-n$ junctions larger than L_p) do not take part in the conduction current, since they recombine before reaching the region of the $p-n$junction. The diffusion length is usually small. For example, it is 10^{-2}cm in the order of magnitude for germanium.

Since the intensity of the current created by the diffusion fluxes of the majority carriers depends on the height $\varphi_0 = eU_c$ of the potential barrier, while the intensity of the conduction current created by the fluxes of the minority carriers is independent of the potential barrier height and since these currents have opposite directions, the total current through the $p-n$ junction is zero only at a certain height of the potential barrier, for which the diffusion current is equal in magnitude to the conduction current.

4.7.6. $p-n$ **Junction as a Barrier layer.** The contact electric field pushes the mobile charge carriers from the region of the $p-n$ junction to regions deeper in semiconductors in contact, where these carriers are majority carriers: electrons are pushed into the n—type semiconductor and holes are pushed into the p—type semiconductor. This effect of the contact field prevents the mobile carriers from penetrating the region of

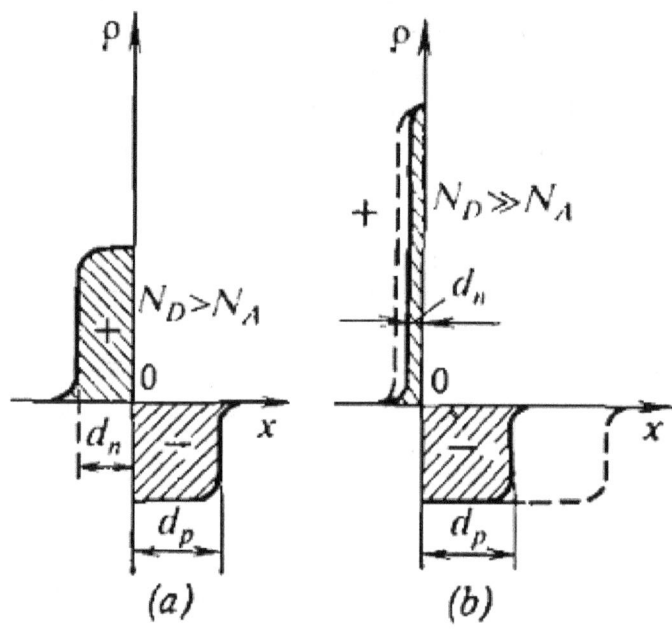

FIGURE 24

the $p-n$ junction from the bulk of semiconductors and making good the shortfall of majority carriers in this region. The uncompensated and rigidly fixed to the lattice donor ions which are left after the departure of electrons create a positive volume charge in the n-type semiconductor region adjoining the junction, while holes leaving the p-type region through the junction create a negative charge of uncompensated ionized acceptor atoms equal in magnitude to the former volume charge. It is these fixed charges that create the contact field and the contact potential difference.

Thicknesses of the depletion layers (d_n on the side of the n-type semiconductor and d_p on the side of the p-type semiconductor) depend on the doping level of each of the regions in contact. In the case of a symmetric $p-n$ junction (when $N_D = N_A$), thicknesses d_n and d_p are equal (see Fig. 64.d). On the other hand, for an asymmetric $p-n$ junction, $d_n \neq d_p$. For example, for $N_D \gg N_A$ the thickness d_n of the depletion layer on the side of the n-type semiconductor is less than the thickness d_p of the depletion layer of the p-type semiconductor (Fig. 66.a). This can be explained by the fact that in a semiconductor with the higher doping level each mono-layer contains more impurity centers, and hence more mobile carriers than a mono-atomic layer of the semiconductor with the lower (loping level. In this case, the $p-n$junction is said lo penetrate deeper into a weakly doped (high-resistance) region of the semiconductor. On the other hand, when $N_D \gg N_A$, the $p-n$ junction is practically localized in the high-resistance p–type region, embracing only a very thin layer of the highly doped n–type semiconductor directly adjoining the $p-n$ interface (Fig. 66.b).

The total thickness of the $p-n$ junction is very small. For example, at a moderate doping level of the p–type and n–type regions in a germanium crystal, the thickness of the junction amounts to about $1\,\mu m$..

4.8. Rectifying Effect of the $p-n$ Junction

4.8.1. Reverse Current. Let us clarify how the application of the external potential difference or, as it is often called, a bias voltage, to the $p-n$ junction changes the conditions for charge carrier transfer through it.

Suppose that an external power source is connected to the $p-n$ junction in such a way that positive terminal is connected to the n–type region and the negative terminal to the p–type region. This polarity of the external source ensures that an additional electric field is created in the $p-n$ junction region, whose direction coincides with that of the contact electric field. As in the case of the metal-to-semiconductor contact, the bias voltage applied in this way is called back (reverse) voltage. Since the junction region is depleted in mobile carriers and has a much higher resistance than the remaining semiconductor, the external potential drop will almost entirely occur in the barrier layer so that the voltage drop in the other regions can be ignored. Hence, the reverse bias U adds up to the contact potential difference U_c, thus increasing the potential barrier in the region of the $p-n$ junction by eU in comparison with its equilibrium value (Fig. 67.b), viz.

$$\varphi' = eU_c + eU = \varphi_0 + eU$$

A comparison of Figs. 67.a and 67.b shows that the external potential difference U shifts the energy levels in the contact regions by eU. An increase in the potential of the n–type region lowers its energy levels, while a decrease in the potential of the p–type region raises the corresponding levels. Naturally, the Fermi levels are also displaced. The difference in the positions of the Fermi levels in the contact regions that arises after the bias voltage is applied (Fig. 67.b) indicates that the equilibrium existing before the application of the bias voltage is disturbed (Fig. 67.a).

The increase in the potential barrier height by eU leads to a decrease in the electron and hole components of the diffusion current through the

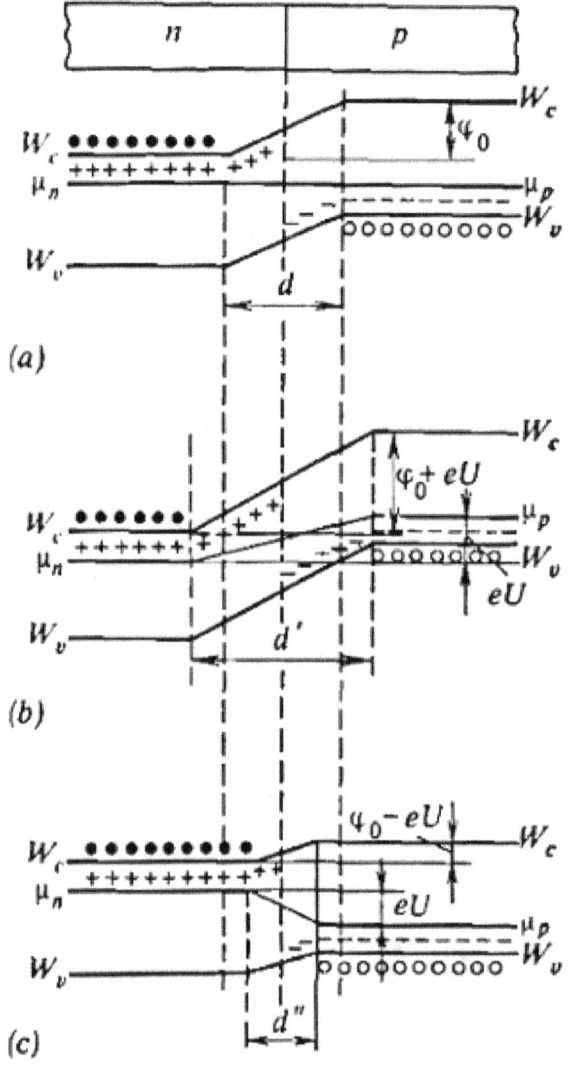

FIGURE 25

junction; the higher the reverse bias, the higher the potential barrier and the smaller the number of the majority carriers capable of surmounting it.

At a certain high bias voltage, the diffusion current through the junction disappears completely.

Along with a decrease in the diffusion current, an increase in the thickness and resistance of the depletion layer at the interface is observed, since as the intensity of the resultant electric field increases (in comparison with that of the contact electric field), the effect of pushing the majority carriers out of the $p-n$ junction region becomes stronger.

The bias voltage practically does not affect the conduction current created by the minority carrier fluxes; the increased electric field in the $p-n$ junction region increases the rate of transfer of the minority carriers through the interface, without changing their number. An increase in the depletion region thickness also does not influence the conduction current, since both the bias field and the contact field promote the transition of the minority carriers through this layer.

Thus, the reverse bias limits the majority carriers flux so that the diffusion current cannot compensate the conduction current. At large reverse bias, the diffusion current tends to zero, and the total current through the junction is determined in practice by the conduction current created by the minority carriers. Since the limiting value of the conduction current does not depend on the bias voltage applied across the junction, it is called the saturation current and denoted by I_s. Sometimes, this current is called the thermal uncontrolled current, which reflects its physical meaning more precisely.

It should be noted that a symmetric broadening of the junction region on both sides of the interface is only observed in the case of a symmetric junction. If, however, the doping level of one of the regions is higher than that of the other ($N_D > N_A$), the depletion region will be mainly broadened in the semiconductor with the lower doping level. If the difference between the doping levels is large, for example, if $N_D \gg N_A$, then almost all the increase in the depletion layer thickness occurs within the weakly doped region (in Fig. 66.b, these changes in the thickness of the layers are shown by dashed lines).

4.8.2. Forward Current. A forward bias applied across the $p-n$ junction (the positive terminal of the external source is connected to the p-type region and the negative terminal to the n-type region) causes a much larger disturbance in the equilibrium. In this case, the shift in the energy levels of the semiconductors in contact reduces the potential barrier (Fig. 67.c). The external voltage U, being subtracted from the contact potential difference, lowers the potential barrier to

$$\varphi'' = \varphi_0 - eU$$

This is accompanied by an increase in the number of electrons capable of surmounting the barrier (the closer the levels are to the bottom of the conduction band, the higher the population density of the energy levels). This increase in the number of the majority carriers overcoming the barrier causes an increase in the diffusion current.

Moreover, the bias field and the contact field in the junction region are directed oppositely. Hence, the resultant field becomes weaker, the depletion region narrower ($d'' < d$), and the resistance of the $p - n$ junction lower, this also causes an increase in the diffusion current.

Since the intensity of the conduction current (which also remains unchanged in the case of a forward bias) is low, it can be ignored when the diffusion current increases with the bias voltage. Thus, in the case of a forward bias, the current flowing through the $p - n$ junction is almost completely determined by the fluxes of the majority carriers, i.e. it is a diffusion current. The majority carriers which have surmounted the potential barrier and got into the neighboring region become the minority carriers in this region, thus increasing the concentration of the minority carriers in the contact region. The difference in concentrations formed causes the diffusion of excess minority carriers from the $p - n$ junction into the bulk of the semiconductor, where they soon recombine. The higher the forward bias, the lower the potential barrier, and the larger the excess concentration of the minority carriers in the $p-n$ junction region. Consequently, as the forward bias increases, the rates of diffusion and recombination grow, and the current flowing through the junction increases. When the bias voltage becomes higher than the contact potential difference, the potential barrier disappears completely. At the same time, the depletion region also disappears and the voltage $U - U_c$ turns out to be distributed over the entire sample. A further increase in the forward potential difference will cause, in accordance with Ohm's law, an increase in current:

$$I = \frac{U - U_c}{R}$$

where R is the resistance of the entire sample.

4.8.3. Injection of Carriers. The penetration of the majority carriers through the $p - n$ junction, caused by a forward bias and accompanied by an increase in the minority carrier concentration in the contact regions, is called the injection of the minority carriers. The ratio of the excess concentration of the minority carriers in the immediate proximity of the $p - n$ junction to the equilibrium concentration of the majority carriers is called the injection level.

In the case of a symmetric $p - n$ junction, the number of electrons $\triangle n_p$, injected from the n-type region to the p-type region is equal to the number $\triangle p_n$ of holes injected from the p-type to the n-type region. For an asymmetric junction, the number of carriers injected from the region with the higher doping level will exceed the opposite flux, and the ratio of the excess concentrations of carriers injected into appropriate regions will be determined by the ratio of the majority carrier concentrations:

$$\frac{\Delta n_p}{\Delta p_n} = \frac{n_{n_o}}{p_{p_o}}$$

If, for example, the concentration of the impurity atoms in the n-type region is 1000 times higher than that in the p-type region, the electron flux from tho n-type to the p-type region will be 1000 times larger than the opposite hole flux. Since the lowering of the potential barrier caused by a forward bias equally facilitates the transition of holes from the p-type to the n-type region and the transition of electrons from the n-type to the p-type region, the additional flux of carriers from the region with the higher concentration will be larger .

For $n_{n_o} \gg p_{p_o}$, the flux of holes injected into the n-type region from the p-type region is negligibly small in comparison with the opposite electron flux, and we can assume that the total diffusion current flowing through the $p-n$ junction is determined by its electron component. The region of a semiconductor from which injection mainly occurs is called the emitter region or simply the emitter, while the region to which injection is predominantly directed is called the base region, or simply the base. In the last example, the n-type region is the emitter, while the p-type is the base.

For an asymmetric $p-n$ junction, both forward and reverse currents are practically created by only one type of carriers. For example, for $n_{n_o} \gg p_{p_o}$," the forward as well as the reverse currents are mainly created by electrons, since in the highly doped n-type region the number of holes (minority carriers) is small, while in the weakly doped p-type region the number of electrons is much larger.

4.8.4. Current-Voltage Characteristic of the $p-n$ Junction. The I vs. U curve (Fig. 68) for the $p-n$ junction has the same form as in the case of a metal-semiconductor junction. For a forward bias, the total current I_f, through the junction, equal to the difference between the diffusion current I_d and the conduction current I_c,

$$I_f = I_d - I_c$$

rapidly increases with the bias voltage due to the increase in the diffusion current at a constant conduction current. At forward bias exceeding $0.1\,V$, when the conduction current can be ignored, the current intensity increases almost exponentially. The essential non-linearity of the characteristic in the initial region of the forward current is explained by a decrease in the thickness of the $p-n$ junction and its resistance with increasing forward bias in the region of its low values. When the forward bias attains several tenths of a volt or more, the characteristic becomes practically linear, since in this region the potential barrier φ (as well as the depletion region) vanish, and the applied potential difference turns out to be distributed over the entire length of the sample (Ohm's law).

The reverse current through the junction is much smaller than the forward current. For this reason, another scale is used for plotting the reverse branch of the current-voltage characteristic: two or three orders of magnitude higher for current and two or three orders lower for voltage (this leads to a kink at the origin which is absent when the same scale is used). The reverse current

$$I_r = I_c - I_d$$

grows quite rapidly in the initial region due to the sharp decrease in the diffusion current with increasing potential barrier height. However, even at reverse bias of the order of $0.1 - 0.2\,V$, this growth ceases. At higher voltages, the diffusion current drops almost to zero, and the reverse current becomes equal to the conduction current which, as we showed above, is almost independent of the applied voltage. A certain increase in the reverse current in the "saturation" region can be explained by the heating of the p-n junction by the current itself and by other side effects.

The rectifier properties of the $p-n$ junction are characterized by the rectification factor. which is defined as the ratio I_f/I_r, between the forward and reverse currents (for similar absolute values of the bias voltages). Usually, the rectification factor ranges between 10^5 and 10^7. This means that the $p-n$ junction has practically unipolar conductivity.

4.8.5. Effect of Temperature on the Rectifier Properties of the $p-n$ Junction. At room temperature most semiconductors, including germanium and silicon, are in the "depletion" state of impurities, since all the impurity centers are ionized. In this state, the n-type semiconductors have the electronic conductivity and the p-type semiconductors have the hole conductivity. When a semiconductor is heated to temperatures close to the intrinsic conductivity temperature T_i, the excitation of the intrinsic atoms of the semiconductor becomes more and more intense. This is accompanied by the simultaneous penetration of electrons and holes. At $T > T_i$, the concentration n_i of intrinsic carriers exceeds the concentration of the "impurity" carriers contained in the semiconductor due to the ionization of impurity centers. In this temperature region, the semiconductor loses the properties of the impurity semiconductor and

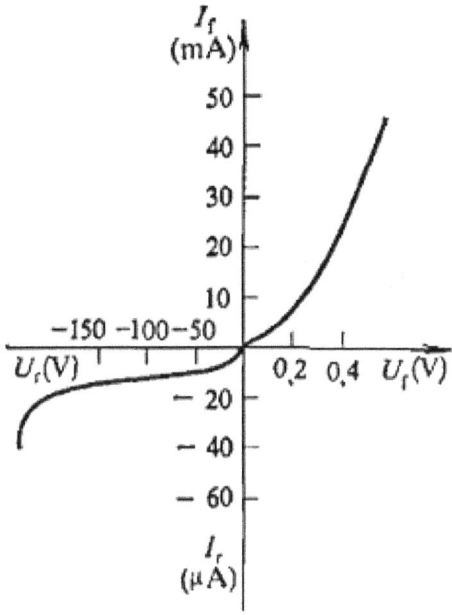

FIGURE 26

becomes an intrinsic semiconductor. In this case, its Fermi level shifts to the middle of the forbidden gap, the potential barrier and the barrier layer disappear, and the $p-n$ junction loses its rectifier properties.

The temperature T_i of intrinsic conductivity depends on the forbidden band width W_g of a semiconductor: the larger W_g, the higher T_i. Therefore, the temperature limit of operation for $p-n$ junctions is determined by the properties of materials from which they are manufactured. For example, in germanium whose forbidden bandwidth $W_g = 0.72\,eV$, the $p-n$ junction can work up to about $75^\circ C$, while in silicon for which $W_g = 1.12\,eV$, the working temperature may reach $150^\circ C$.

4.9. Breakdown of the $p-n$ Junction

If we gradually increase the reverse bias voltage, upon attaining a certain value U_{br}(Fig. 69) a sharp increase will be observed in the current flowing through a circuit containing the $p-n$ junction and no limiting resistor. This is called the breakdown of the $p-n$ junction. The breakdown can be electrical and thermal in nature. The electrical breakdown does not destroy the junction, and if it is not followed by a thermal breakdown, the properties of the $p-n$ junction recover after the reverse voltage is removed. There are two types of electrical breakdown: avalanche and tunnel

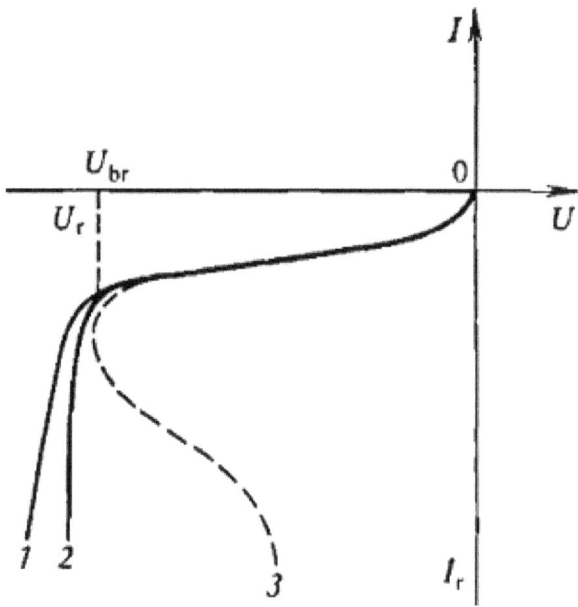

FIGURE 27

4.9.1. Avalanche Breakdown. This type of breakdown is associated with the effect of the avalanche multiplication of carriers in the $p-n$ junction. At a certain value of reverse bias voltage, the field intensity in the $p-n$ junction becomes so high that the minority carriers accelerated by the field acquire enough energy to ionize neutral atoms of the semiconductor in the junction region. As a result of the ionization, the number of carriers creating the reverse current grows abruptly. The avalanche breakdown is usually observed at a reverse bias of the order of tens or hundreds of volts. It is typical of quite thick $p-n$ junctions, in which each minority carrier accelerated by the electric field causes multiple ionization of atoms in the depletion layer. These junctions usually exhibit the tunnel breakdown.

4.9.2. Tunnel Breakdown. This phenomenon originates from the tunnel effect which is observed due to the direct influence of a strong electric field on the atoms of the semiconductor crystal lattice in the $p-n$ junction. Under the influence of this field, the valence bond is ruptured and an electron becomes a free carrier, moving into the interstitial space and leaving a hole instead. The energy band diagram of the tunnel breakdown is shown in Fig. 70. Electrons from the valence band of the p-type semiconductor move into the conduction band of the n-type semiconductor without changing their energy, crossing the forbidden band of the $p-n$ junction. The necessary condition for tunneling

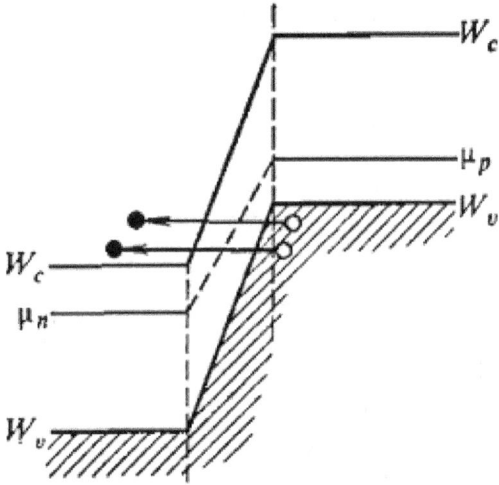

FIGURE 28

is the existence of a vacant level in the conduction band of the n–type semiconductor whose energy corresponds to the energy of the electron arriving from the p–type side.

The tunnel breakdown is observed in thin $p-n$ junctions which can be formed only at the interface between highly doped regions. In order to provoke the tunnel breakdown, a field of $10^5 - 10^6\,V/cm$ is required. Since the tunnel break- down only occurs in thin junctions of the order of $10^{-5} - 10^{-6}\,cm$, the reverse potential difference of only several volts is sufficient for obtaining the breakdown values of field intensity.

The rise in current during the tunnel breakdown (curve 2 in Fig. 69) is even steeper than that during the avalanche breakdown (curve 1). As was mentioned above, the properties of the $p-n$ junction recover after the electric breakdown (either avalanche or tunnel) when the reverse bias is removed. For this reason, $p-n$ junctions are frequently used in technology under the breakdown conditions (crystal stabilizers, or Zener diodes, tunnel backward diodes, etc.).

4.9.3. Thermal Breakdown. If the amount of heat liberated in the $p-n$ junction exceeds the amount of heat removed from it, the heating of the junction intensifies the carrier generation and hence leads to an increase in the current flowing through the junction. This, in turn, results in a further increase in temperature, and so on. As a result of such avalanche-type overheating, the current intensity continues to grow even if the voltage decreases (curve 3 in Fig. 69), causing the destruction of the semiconductor material. Thermal breakdown can be either self-induced or may be a consequence of the developing electric breakdown. For this reason, a limiting resistor is usually connected in series with a $p-n$ junction, the resistance being chosen in such a way that the current cannot exceed a certain value.

4.9.4. Surface Breakdown. The avalanche or tunnel electric breakdown of the $p-n$ junction may occur not only in the bulk of a semiconductor but also on its surface. The surface breakdown can be induced by a distortion of the electric field in the $p-n$ junction due to surface charges. A surface charge may appear as a result of the destruction of the crystal lattice or due to defects and impurities present in it (particularly, adsorbed water molecules). In some cases, a surface charge narrows the barrier layer at the surface and increases the field intensity in the surface layer, due to which the breakdown at the surface appears at lower values of the reverse bias voltage than in the bulk. In order to reduce the probability of a surface breakdown, the surface of semiconductors is coated with materials to protect the $p-n$ junction from moisture and other active impurities.

4.10. The Electric Capacitance of the $p-n$ Junction

When a bias voltage is applied across the $p-n$ junction, it exhibits the properties of a capacitor.

The $p-n$ junction is the region in a semiconductor, depleted in mobile carriers and similar in its properties to a dielectric layer in which fixed unlike electric charges are distributed over a volume on both sides of a certain plane. In a semiconductor, these are charges of ionized impurity atoms: the n-type region of the junction is charged positively, while the p-type region is charged negatively. This structure of the $p-n$ junction allows us to identify it, to a certain extent, with a plane-parallel capacitor. The application of the reverse bias leads to a further pushing of mobile carriers out of the boundary regions, and increases the thickness of the junction and the number of uncompensated fixed impurity ions on both sides of the interface. Hence, the $p-n$ junction responds to a change in the reverse bias by ΔU by a ΔQ increase in the charge, and this is what a capacitor does. The quantity

$$C_b = \Delta Q/U$$

is called the barrier capacitance. Sometimes, it is called the charge capacitance or simply the capacitance of the junction.

As in the case of capacitors, the barrier capacitance depends on the area of the $p-n$ junction, the dielectric constant of the depletion layer, and its thickness. In most cases, the area of the $p-n$ junction is small,

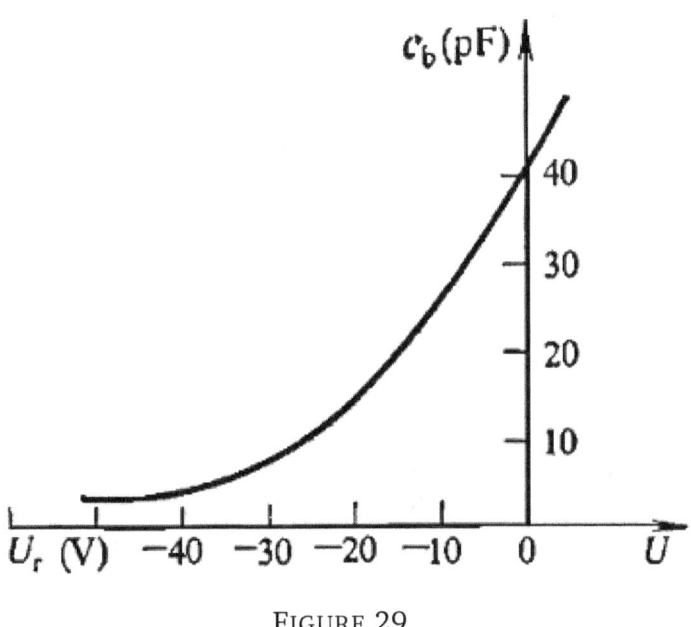

FIGURE 29

nevertheless, the barrier capacitance is large due to the thinness of the depletion region. By varying the thickness of the $p-n$ junction, it is possible to change the barrier capacitance from several to dozens of thousands of pic-farads per square centimeter. The most important property of the barrier capacitance is its dependence on the bias voltage applied across the junction. The thickness of the barrier layer increases with the reverse bias, which results in a decrease in the capacitance. Figure 71 illustrates the variation of the barrier capacitance as a function of the applied reverse voltage. This relationship is used in a special type of semiconductor diodes called varicaps (variable capacitor). Varicaps are employed as variable capacitors for tuning the frequency of oscillatory circuits. In rectifier diodes, the barrier capacitance is harmful since it by-passes the $p-n$ junction by conducting some alternating current in a "roundabout" way (especially in the high-frequency region). The barrier capacitance also appears with a forward bias if $U < U_c$ (at such voltages, a barrier layer still exists).

CHAPTER 5

Semiconductor Devices

5.1. Hall Effect and Hall Pickups

The Hall effect is an interesting phenomenon. But besides a purely physical importance, it is very attractive from a practical point of view. This is due to wide applications of the Hall effect for determining electrical properties of various solids, including semiconductors. This effect can be used for determining the sign of the majority charge carriers, their concentration and mobility, i.e. the parameters that determine the basic electro-physical properties of a semiconductor. The analysis of specific semiconductor devices and setups would be impossible without the information about, these parameters. For this reason, before considering the operating principles of semiconductor devices, we shall give an idea of the Hall effect.

5.1.1. Lorentz Force. If a unit positive charge $+e$ (a hole) moves at a velocity v in a magnetic field with an induction B, the force of the magnetic field acting on this charge is given by

$$F_{mang} = evB \sin(B, v)$$

When the velocity vector v is perpendicular to the magnetic field vector B, $\sin(B, v) = 1$, and the expression for F_{magn} becomes

$$F_{mang} = evB$$

The direction of the force F_{mang} can be determined with the help of the left-hand rule. If we also take into account the force $F_{el} = eE$ acting on a charge in an electric field, the resultant force acting on the charge moving in the electromagnetic field with $v \perp B$ will be given by

$$F = e(E + vB)$$

This expression was obtained for the first time in the vector form by the famous Dutch scientist H. Lorentz and is called the Lorentz force. The quantity F_{magn} is often called the Lorentz force. This quantity does not have a special name and it would be more correct to call it the magnetic component of the Lorentz force, or simply the magnetic force.

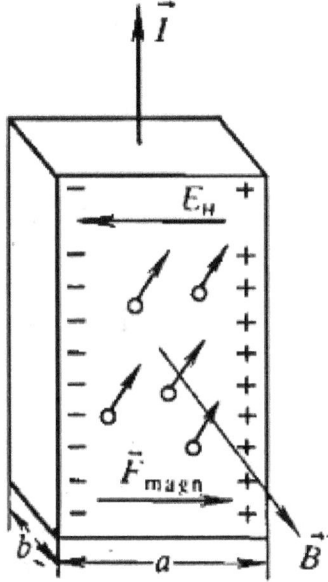

FIGURE 1

Having got the idea about the Lorentz force, let us now consider the Hall effect.

5.1.2. Hall Effect in the Extrinsic Semiconductors. Let us place a sample of a current-carrying $p-$ type semiconductor into a magnetic field directed perpendicularly to the current (Fig. 72). According to the left-hand rule, holes will deviate under the action of the magnetic force F_{mang} to the right face of the crystal. As a result of the accumulation of holes, this face will become positively charged, while the left face will acquire an uncompensated negative charge due to the departure of holes. Such a spatial charge distribution in the crystal leads to the appearance of a transverse potential difference between the sample faces, which is essentially the Hall effect.

The process of the accumulation of charges will last until the increasing force of the transverse electric field balances the magnetic force. Once the balance is established, the charge accumulation ceases and the crystal reaches the stationary state in which holes ; move parallel to the lateral faces. The potential difference established between the oppositely charged faces is called the Hall e.m.f .and is denoted by V_H. The intensity of the electric field created in the crystal can be expressed as $E_H = V_H/a$, where a is the distance between the oppositely charged faces. The force of the field acting on holes is

$$F_{el..H} = eE_H$$

Using the condition that forces acting on holes in the transverse direction are balanced in the stationary state, we can write

$$F_{mang} = F_{el.H} \text{ or } evB = eE_H$$

hence we obtain the expressions

$$E_H = vB$$

for the hall field and

$$V_H = E_H a = vBa$$

for the Hall e.m.f.

The current flowing through the sample is measured experimentally instead of the velocity v of the holes. The magnitude of this current is $I = abj$, where ab is the cross-sectional area of the sample and j is the current density. Using the expression $j = nev$ which we obtained earlier and replacing n by p (since we have p–type conductivity), we can express the total current through the sample as

$$I = abj = abpev$$

Hence we can determine the velocity v:

$$v = \frac{1}{pe} \cdot \frac{I}{ab}$$

Substituting its value into the expression for the Hall e.m.f., we obtain

$$V_H = \frac{1}{pe} \cdot \frac{BI}{b}$$

The proportionality factor that relates the Hall e.m.f. to the magnetic field intensity and to the current flowing through the sample is called the Hall constant and is denoted by

$$R_H = \frac{1}{pe}$$

Using this notation in the expression for the Hall e.m.f. and solving it for the Hall constant, we obtain the following expression:

$$R_H = \frac{V_H b}{BI}$$

5.1.3. Practical Application of the Hall Effect. If we know the value of the magnetic induction B, current I, and the dimension b of the sample in the direction of the applied magnetic field, we can determine the value of the Hall constant by measuring the Hall e.m.f. created in the sample and substituting it into the last formula. Having determined the numerical value and the sign of the Hall constant, we can find the concentration and sign of the majority carriers in the material under investigation. It can be seen that the sign of the Hall constant is positive for p–type semiconductors and negative for n–type semiconductors.

If we measure the electrical conductivity $\sigma = peu$ of the sample during the Hall effect, it is possible to determine the mobility u of the majority carriers:

$$u = \frac{1}{pe}.\sigma = R_H \sigma$$

Thus, the simultaneous determination of the Hall constant and the electrical conductivity of the sample makes it possible to obtain the important information about the majority carriers of the material under investigation, viz. their sign, concentration and mobility. For this reason, the Hall effect is widely used as one of the basic methods of determining the elector-physical properties of semiconductors and metals.

It is interesting to note that the Hall effect in such typical metals as zinc, beryllium or cadmium resulted in positive values of the Hall constant rather than negative as was expected. The conduction band of these metals proved to be almost completely filled and only a narrow region near its top is vacant. This is typical of semiconductor crystals of p–type.

By analyzing the temperature dependence of the Hall constant in semiconductors, it is possible to determine the temperature dependence of the carrier concentration. If we simultaneously register the variation of electrical conductivity, we can determine the temperature dependence of the carrier mobility as well. The temperature dependence of the Hall constant can be used to estimate the depth of the impurity energy levels (i.e. the separation between the impurity levels W_d or W_a and the corresponding bands), as well as the forbidden band width W_g. Figure 73 shows for comparison the dependence $\ln n = f(1/T)$ which we have already analyzed and the curve $\ln R_H = f(1/T)$. It can be seen from the figure that the slope of the segment cd for extrinsic conductivity in the low-temperature region ($T < T_s$) can be used for determining the activation energy of the donor level W_d, while the slope of the segment ab of the intrinsic conductivity (for $T > T_i$) determines the forbidden band width W_g.

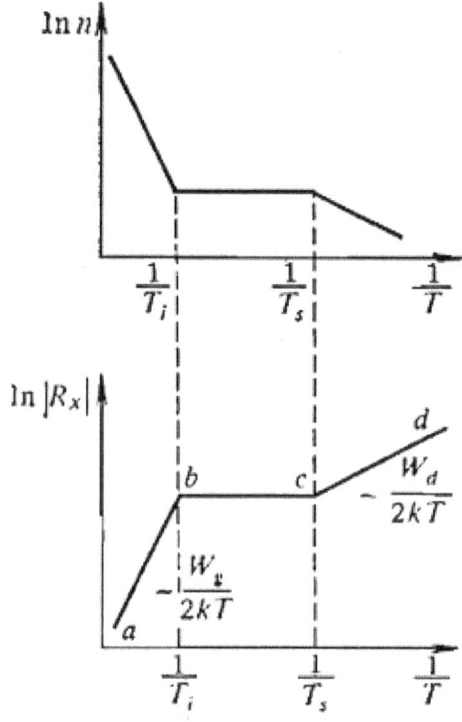

FIGURE 2

It should be noted here that a certain error was made while calculating the Hall constant. When considering the motion of holes, we took into account only the velocity due to the action of the electric field, i.e. the drift velocity, and ignored the velocity of random thermal motion. If we take into consideration the thermal motion and a certain velocity distribution of carriers, we obtain a different expression for the Hall constant: :

$$R_H = \frac{A}{pe}$$

The constant A is determined by the mechanism of carrier scattering in crystals and has different values for different cases depending on the type of crystal lattice and the operating temperature region.

5.1.4. The Hall Effect in Semiconductors with Mixed Conductivity. If the conductivity of a semiconductor is neither of the n-type nor of the p-type and both electrons and holes are responsible for electrical conductivity (mixed conductivity) the Hall effect has a more complicated form. This is due to the fact that the magnetic field deviates electrons and holes in the same direction. Indeed, the action of the magnetic field on holes is determined by the formula $F_{mang} = evB$. To apply this formula for electrons, we must change the sign of the charge e as well as of the velocity v, thus the sign, and the direction of the magnetic component F_{mang} of the Lorentz force will not change. For this reason, in the Hall effect for crystals with mixed conductivity, electrons and holes move towards the same lateral face. The sign of the charge of a certain face will be determined by the ratio of concentrations and mobilities of carriers. If the electron concentration is higher (electron mobility is normally higher than that of holes) negative charges will be accumulated on the face towards which the carriers are deviated, while the opposite face will acquire an uncompensated positive charge. The transverse Hall field created in this case prevents new electrons from arriving at this region and at the same time accelerates the motion of arriving holes. When the electron and hole fluxes deviated by the magnetic field become equal, the process of charge accumulation ceases, and a stationary state sets in, which is characterized by a certain value of the Hall e.m.f. The value of the Hall constant in this case in defined by the expression:

$$R_H = \frac{A}{e} \frac{u_p^2 p - u_n^2 n}{(u_p p + u_n n)^2}$$

As before, n and p denote the electron and hole concentrations, and u_n and u_p their mobilities.

5.1.5. Hall Effect in Intrinsic Semiconductors. In an intrinsic semiconductor, the charge carrier concentrations are equal ($n = p$). Hence, the last expression is reduced to :

$$R_H = \frac{A}{n_e} \cdot \frac{u_p - u_n}{u_p + u_n}$$

It follows from this expression that the Hall effect in an intrinsic semiconductor and the sign of the Hall constant are determined by the ratio of carrier mobilities. Normally, the mobility of electron. is higher than the mobility of holes, and hence the Hall constant is not large and has the minus sign (the Hall effect is said to be weak).

If the mobilities of electrons and holes in an intrinsic semiconductor are equal ($u_n = u_p$), the Hall constant is equal to zero, i.e. the Hall e.m.f. does not appear. Although the Hall effect is not observed in such semiconductors, it does not mean that charge carriers creating the current are not deflected by an external magnetic field.. n this case the fluxes of deflected electrons and holes are equal, consequently, the charges are compensated completely, and the Hall e.m.f. does not appear. The action of the magnetic field is reduced to a redistribution of carrier through the sample: the concentration is higher at the face towards which the carriers are displaced in comparison with the opposite face. The local electrical conductivity along lateral faces of the crystal changes accordingly.

5.1.6. Hall Pickups-and Their Applications. Tho practical application of the Hall effect is not limited to the study of electro-physical properties of semiconductors and metals. The field of application of the Hall effect for various purposes is so broad that we can consider it the most widely used effect in science. However, in spite of the large variety of the spheres of its application and measuring instruments where it is used, the main element in all cases is the so called Hall pickup (Fig. 74). It consists of a thin rectangular plate cut from a semiconductor crystal with a high carrier mobility. Current-carrying electrodes 1 are soldered to two opposite lateral faces of this plate, and the Hall electrodes 2 are soldered to the remaining two lateral faces, which are used for registering the Hall e.m.f. appearing in the plate (Fig. 74.a). A semiconductor film is often used instead of a crystal plate, the former is deposited by the evaporation method or is sprayed onto a thin insulator substrate usually taken in the form of a thin mica sheet (Fig. 74.b)

5.1.7. Hall Magnetometers. The devices based on the Hall effect are used most of all for measuring the magnetic field intensity. These instruments are called magnetometers and consist of a Hall pickup and a potentiometer. It was shown above that the Hall e.m.f. is proportional to the magnetic field intensity. Consequently, knowing the parameters of the Hall pickup and the applied current, we can estimate the magnetic field intensity from the measured value of the Hall e.m.f. Magnetic fields frequently have to be measured in very narrow gaps. Hall pickups in the form of a thin semiconductor films deposited onto thin mica substrates are very suitable for this purpose.

The Hall effect can be used both to measure permanent magnetic fields and to investigate the variable magnetic fields as well as the dynamics of the variation of magnetic fields. This can be done because of very small inertia of the effect. Hall pickups can be used for measuring variable magnetic fields up to a frequency of 10^{12} Hz. Extremely small sizes of Hall pickups (they can be manufactured in the form of plates $0.5x0.5$ mm in size) make them applicable for investigating the magnetic field topography, i.e. for determining the non-uniformity of the field distribution in space or in a plane.

5.1.8. Heavy-Current Ammeters. Hall ammeters are another example of the application of the Hall effect for measuring purposes. They

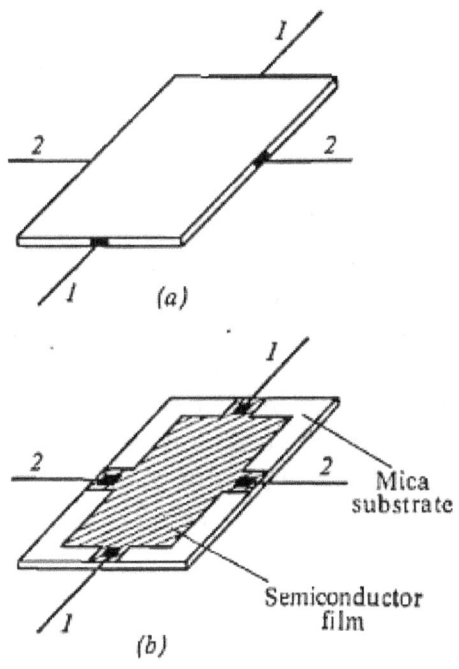

FIGURE 3

were designed for extremely heavy currents, when conventional ammeters cannot be used.In order to create the primary direct current in such a device, a stabilized source of d.c. voltage is used, which can be a battery. The current being measured is used for creating the magnetic field. Since the Hall e.m.f. appearing in the pickup is proportional to the magnetic field, changes in the controlled current responsible for a change in the magnetic field are transformed into the variations of the Hall voltage being measured.

5.1.9. Hall Pickups as Signal Transducers. One more field of wide application of the Hall effect is the transformation of a direct current into alternating current and vice versa. When a d.c. is transformed into a.c., the current-carrying electrodes of the pickup are connected to the circuit of the d. current being transformed and the pickup is placed into the field of an electromagnet. The a.c. signal is supplied to the windings of this magnet. The varying field creates an e.m.f. on the Hall electrodes, whose frequency corresponds to the a.c. frequency. The magnitude of this e.m.f. is proportional to the magnetic field intensity and to the primary current being transformed. Thus, the obtained a.c. signal can be additionally controlled in the process of transformation.

5.1.10. Hall Microphone. This device is an example of a rather peculiar application of the Hall effect for transformation of currents. The microphone pickup is placed in the magnetic field in such a way that its plane is parallel to the magnetic field lines. In this case, the Hall effect does not appear in the pickup in spite of the primary current from an external power supply connected to the current terminals. This is so because the normal component of the magnetic field is equal to zero. Since the microphone pickup is mechanically connected to the membrane, acoustic vibrations are transformed into mechanical vibrations by the membrane and transmitted to the pickup. The vibrations of the pickup in the magnetic field create an alternating e.m.f. across the Hall electrodes, which is synchronized with the acoustic signals arriving at the membrane. The generated alternating voltage is supplied to the input of the amplifier and is then transformed into the radio-frequency oscillations or supplied to a loudspeaker.

There are many other examples of application of the Hall effect in various fields. Among the most interesting cases, we can mention its application for a compass, pressure gauge, instrument for measuring small linear and angular displacements, instrument for measuring transient time difference, various types of amplifiers and modulators, etc.

5.2. Semiconductor Diodes

At present, semiconductor devices are used practically in all branches of electronics and radio-engineering. In spite of the great variety of these devices, they are based, as a rule, on the operation of an ordinary $p-n$ junction or a system of several junctions.

A semiconductor diode contains a single $p-n$ junction each of whose regions is connected through ohmic contacts with metallic terminals.

5.2.1. Rectifier Diodes. Semiconductor diodes are mainly used for rectifying an alternating current

Figure 75 shows the simplest circuit in which a semiconductor diode plays the role of a rectifier. A source of a.c. voltage $u_\sim$ a diode D ,and a load resistor R_1, are connected in series. The forward direction of the diode is indicated by the arrow (from anode to cathode).

Suppose that the voltage across the terminals of the source varies sinusoidally (Fig. 76.a). During a positive half-period, when the "plus" pole is connected to the anode and the "minus" to the cathode, the diode operates in the forward direction, and the current flows through it. The instantaneous value i of the current is determined by the instantaneous voltage u across the terminals of the source and the load resistance (in the forward direction, the resistance of the diode is small and can be

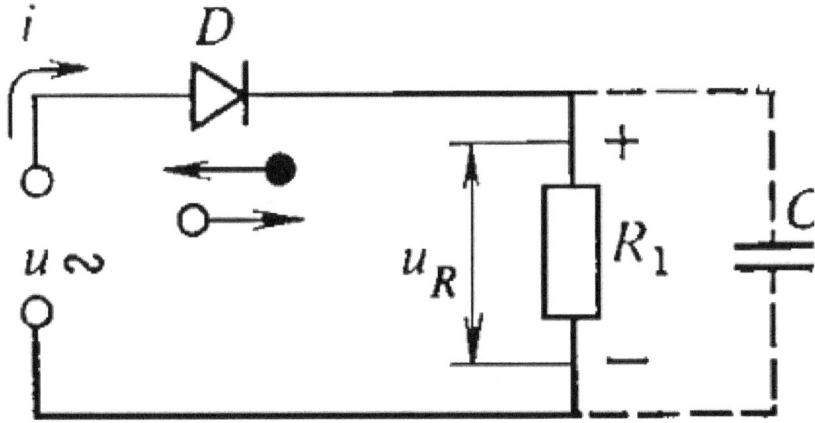

FIGURE 4

neglected). During the negative half-period, almost zero current flows through the diode. Hence, a pulsating current flow through the circuit, as is shown schematically in Fig. 76.b. The voltage u_R across the load resistor will also be pulsating. Since $u_R = iR_l$, the variations in u_R follow the variations in the current i. The voltage across the load resistor always has the same polarity corresponding to the direction of the forward current: the "plus" at end of the resistor connected to the cathode and "minus" at the opposite end.

The circuit we have just considered provides half-wave rectification. In order to decrease pulsations in the rectified voltage, ripple filters are used. The simplest way of smoothing the voltage is to connect a capacitor C in parallel to the load resistor (the dashed part of the circuit in Fig. 75). During a positive half-period, some of the current flowing through the diode is consumed for charging the capacitor. On the other hand, during a negative half-period, when the diode is cut off, the capacitor discharges through R_l, thus maintaining the current in the same direction. As a result, the voltage pulsations in the load resistor are considerably smoothed.

5.2.2. Stabilizer Diodes (Stabilitrons). We showed above that during an electrical breakdown of $p-n$ junction, the current across the junction rises very steeply. This part of the current-voltage characteristic can be used for stabilizing voltage. For this purpose, silicon diodes are mainly used. Figure 77 shows the reverse branch of the current-voltage characteristic of such a diode. It can be seen that when the breakdown voltage is attained, very small variations of voltage cause very large changes in the current Slowing across the diode (tens of times).

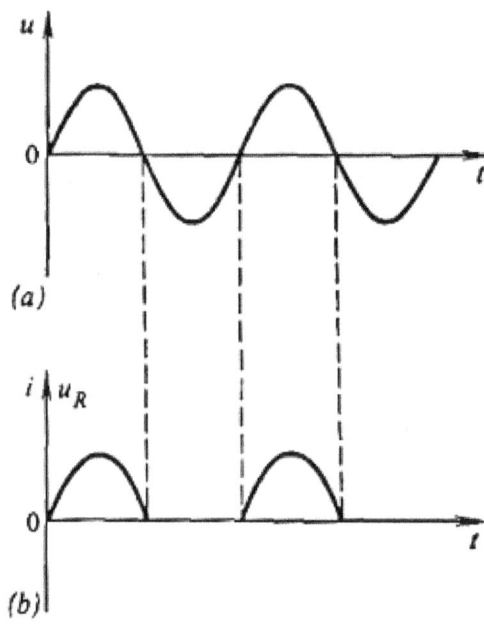

FIGURE 5

Figure 78 shows the diagram of a circuit for stabilizing the voltage. The stabilizer diode is connected in parallel to the load resistor R_l across

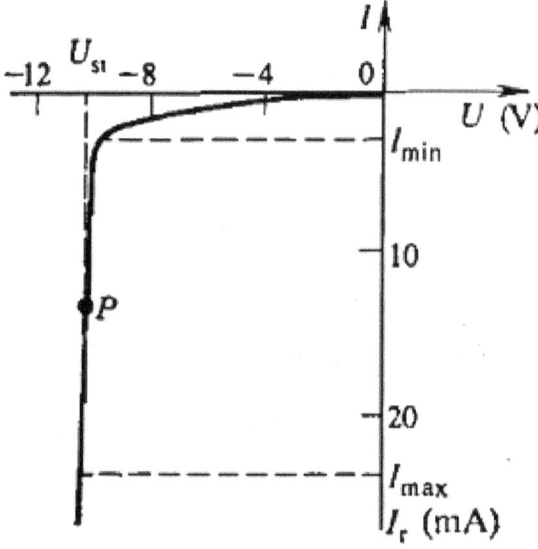

FIGURE 6

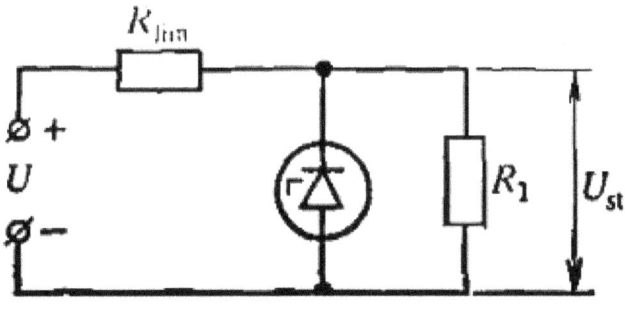

FIGURE 7

which a stabilized voltage must be obtained when there is an alternating voltage across the source terminals. A limiting resistor is series connected to the diode. Its resistance R_{lim} is chosen in such a way that the current corresponding to the middle part of the operating region flows through the stabilizer diode (point P in Fig. 77). If, for example, the voltage across the terminals of the source increases, the current in the circuit will also increase. However, an increase in the current through the stabilizer (even within broad limits) occurs at a practically constant voltage, due to which the voltage across the load R_l remains unchanged. At the same time, an increase in the current through the limiting resistor leads to an increase in the potential drop across it. Hence, all the variations in the source voltage die out in the limiting resistor, while a constant voltage U_{st} is maintained across the load resistor.

At the present time, silicon $p-n$ junctions are used for manufacturing stabilizer diodes that can handle voltages ranging from 1 to 300 V.

5.2.3. Varicaps. Diodes used as variable capacitors (varicaps) also operate in the reverse bias regime. Unlike ordinary variable capacitors with a mechanical control, the capacitance in varicaps is controlled by varying the bias voltage. Figure 79 shows the circuit for retuning the frequency of an oscillatory circuit with the help of a varicap.

When the reverse bias applied to the varicap is varied by a potentiometer R, its barrier capacitance changes, and hence the resonance frequency of the oscillatory circuit also varies. The series resistor R_s, which has a sufficiently high resistance, prevents the oscillatory circuit from being shunted by the potentiometer R, and C_{bl} is the blocking capacitor. If the latter were absent, the varicap would be short-circuited by the coil L (with respect to the constant component of the voltage).

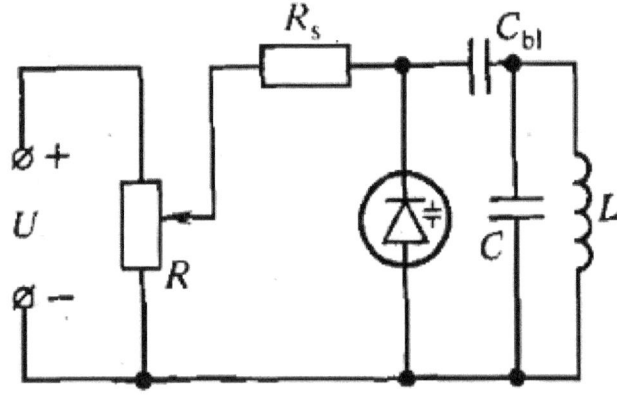

FIGURE 8

5.3. Tunnel Diodes

In the large family of semiconductor devices, there is a group for which an increase in voltage in a certain part of the current-voltage characteristic ($\Delta U > 0$) is accompanied by a decrease, rather than an increase, in the current intensity ($\Delta I < 0$). Such regions of the current-voltage characteristic correspond to a negative resistance:

$$R = \frac{\Delta U}{\Delta I} < 0$$

The most widespread and perhaps the most interesting of all negative resistance devices are the tunnel diodes. The idea of employing the tunnel effect for a semiconductor diode was put forward by the Soviet scientists Frenkel and Ioffe in 1932, but it was only in 1958 that the Japanese engineer Esaki manufactured the tunnel diode.

5.3.1. Manufacturing of the Tunnel Diodes. Like ordinary rectifier diodes, the tunnel diode can be obtained by alloying a piece of metal into a semiconductor wafer, for example, indium into the n–type germanium. In other words, in order to create the tunnel diode, the $p-n$ junction must be obtained. However, unlike the manufacturing of ordinary diodes, for tunnel diode we must use a semiconductor with a very high doping level as the substrate, i.e. it must have a very high impurity concentration. While the doping level of semiconductors used for ordinary diodes does not as a rule exceed $10^{17}\,cm^{-3}$, in semiconductors intended for manufacturing tunnel diodes the doping impurity concentration reaches $10^{19} - 10^{20}\,cm^{-3}$.

5.3.2. The $p-n$ Junction between Degenerate Semiconductors.
It was shown above that semiconductors with such a high doping level are degenerate: their Fermi levels lie in the allowed bands (in the $n-$type degenerate semiconductor, the Fermi level lies in the conduction band, while in the $p-$type degenerate semiconductor, it lies in the valence band). Such an arrangement of the Fermi levels leads to a high contact potential difference at the junction between the degenerate semiconductors, which is almost double the contact potential difference in ordinary diodes. Since the Fermi levels in the tunnel diodes lie beyond the forbidden gap, the potential barrier at the interface of these diodes is always higher than the forbidden band width. Figure 80.a shows the energy band diagram of two highly doped degenerate semiconductors (the $n-$type and $p-$type) before they are placed in contact, while Fig. 80.b gives the energy band diagram after the $p-n$ junction is formed. It can be seen from the latter figure that upon establishing equilibrium between the degenerate $n-$type and $p-$ type regions, the energy bands overlap on the external energy scale: the bottom of the conduction band of the $n-$type semiconductor lies below the top of the valence band of the $p-$type semiconductor. Hence, electrons located, for example, near the Fermi level in the $n-$type and $p-$type regions have the same energy, and the only obstacle for their transition from one region to the other is the forbidden energy band which is a kind of the potential barrier for them.

Another feature of the $p-n$ junction between degenerate semiconductors is its extremely small thickness d, being of the order of $10^{-6}cm$. As a matter of fact, due to a high density of free carriers, their departure even from a thin boundary layer is associated with the formation of a large number of uncompensated charged donor and acceptor impurity centres. These are sufficient for creating an equilibrium potential barrier.

5.3.3. Tunnel Transitions of Electrons in Equilibrium State.
The extremely small thickness of the $p-n$ junction, combined with the overlapping of bands, due to which regions with similar allowed energy are formed on both sides of the junction, create favorable conditions for tunnel transitions: electrons from the conduction band of the $n-$type region move to the valence band of the $p-$type region, and vice versa (see Fig. 80.b). Naturally, the necessary condition for an electron tunneling through the barrier is the presence of vacancies on the side of the barrier to which the electron goes. But we know that the Fermi level is characterized by that its probability of being filled is equal to 1/2. Consequently, there will always be a vacancy behind the barrier for an electron with an energy close to the Fermi energy.

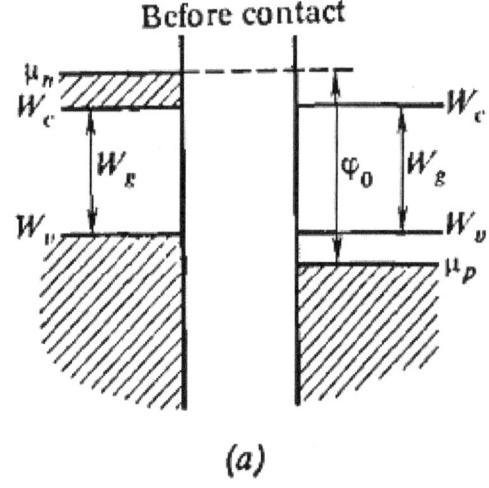

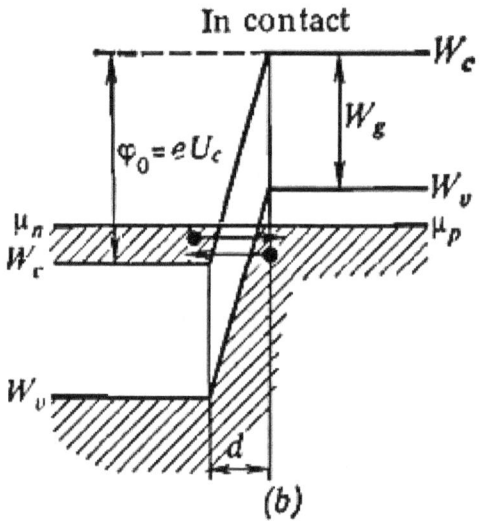

FIGURE 9

In equilibrium (in the absence of the bias voltage), the number of tunnel electron transitions from left to right is equal to the number of opposite transitions from right to left, and the resultant tunnel current is zero. Of course, besides tunnel transitions in the diode under consideration there are also above-the-barrier transitions of the majority and minority carriers, and these create the diffusion and the conduction currents. But, firstly, in equilibrium these currents are equal in magnitude and opposite in direction, so that in total they do not create current. And, secondly, the number of above-the-barrier transitions is negligibly small

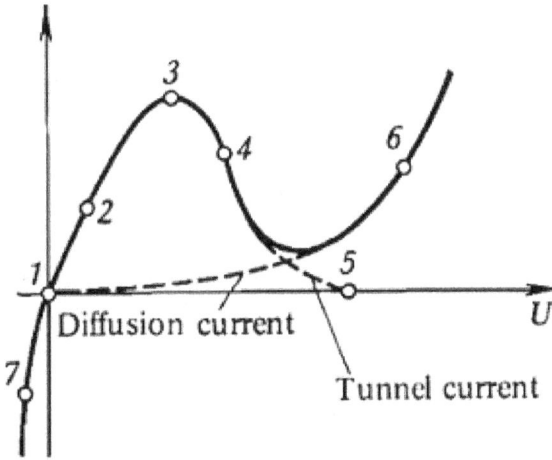

FIGURE 10

in comparison with the number of the tunnel transitions. Thus, in the absence of the external bias voltage, the current through a tunnel diode is zero. This corresponds to the origin of the current-voltage characteristic of the device (point 1 in Fig. 81).

5.3.4. Operation of the Tunnel Diode at Forward Bias Voltage. If a moderate positive bias is applied to the diode, the energy bands will be shifted. As a result, the potential barrier at the interface will become a little lower, and the unfilled part of the valence band of the p-type semiconductor will be opposite to the filled part of the conduction band of the n-type semiconductor (Fig. 82.a). The equilibrium between the tunnel transitions of electrons from left to right and from right to left will be disturbed. Indeed, in the region where the filled parts of the bands overlap, the indicated transitions compensate each other (dashed arrows in the figure), while transitions from the upper region of the filled part of the conduction band of the n-type semiconductor (solid arrow) are not compensated by opposite transitions because the opposite region of the valence band of the p-type semiconductor is empty. The uncompensated electron flux from the n-type semiconductor to the p-type semiconductor forms the forward current through the diode (point 2 on the curve in Fig. 82).

An increase in the positive bias leads to greater overlapping between the filled region of the conduction band of the n-type semiconductor and the empty region of the valence band of the p-type semiconductor, resulting in an increase in the tunnel current through the diode. This current attains its maximum value (point 3 in Fig. 81) when the Fermi level of the n—type semiconductor is opposite to the top of the valence band of the p—type region (Fig. 82.b).

Any further increase in the forward bias is accompanied by a decrease in the overlapping between the filled part of the conduction band of the n—type semiconductor and the empty part of the valence band of the p—type. Hence the conditions for electron transitions from the n—type to the p—type region become less favorable (Fig. 82.c). Electrons from the upper part of the filled region of the conduction band of the n— type semiconductor are now opposite to the forbidden energy band of the p—type semiconductor, so their transition to the p—type region becomes impossible. Thus, we arrive at a paradoxical (at first sight) phenomenon: an increase in the potential difference applied to the device in the forward direction is accompanied by a decrease rather than an increase in the current across it (point 4 in Fig. 81). A falling region corresponding to a negative resistance appears on the current-voltage characteristic of the diode.

The tunnel current will decrease as the forward voltage increases until the bottom of the conduction band of the n—type semiconductor aligns with the top of the valence band of the p—type semiconductor

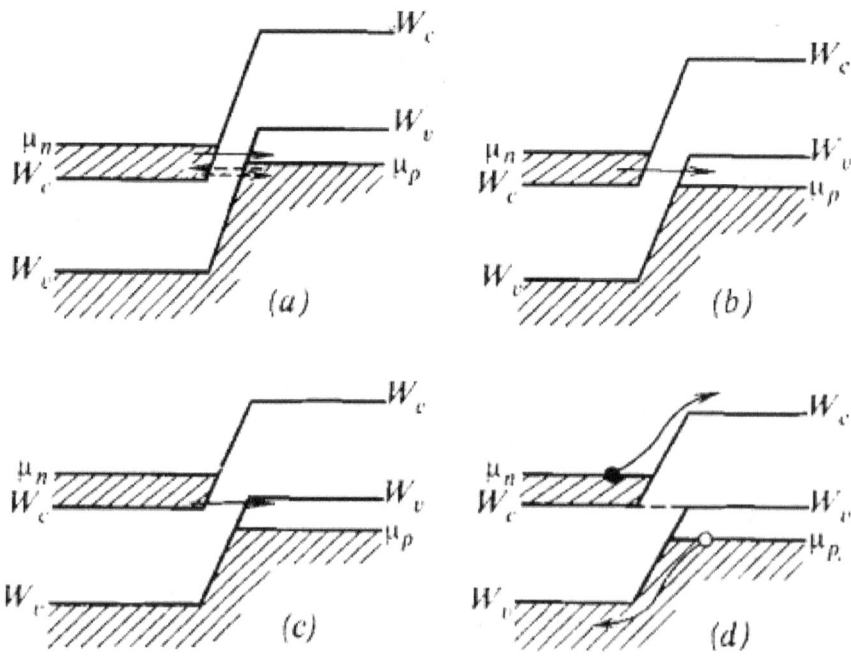

FIGURE 11

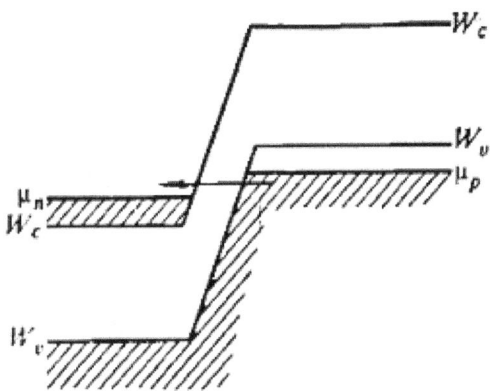

FIGURE 12

(Fig. 82.d). In this arrangement, tunnel transitions become impossible in principle, and the tunnel current drops to zero (point 5 in Fig. 81).

However, as can be seen from the shape of the current-voltage characteristic (see Fig. 81), the current through the diode begins to increase with forward voltage instead of vanishing. This is because a high forward bias voltage considerably decreases the potential barrier at the $p-n$ interface. As a result, the probability of above- the-barrier transition of carriers through the interface increases, i.e. the above-the-barrier injection of electrons from the n—type semiconductor and holes from the p—type semiconductor becomes possible (see Fig. 82.d). As in the case of ordinary diodes, the resulting diffusion current increases with the forward voltage which makes the potential barrier at the $p-n$ interface lower and lower (the ascending region containing point 6 in Fig. 81).

5.3.5. Operation of the Tunnel Diode at the Reverse Bias. When the reverse bias is applied, tunnel transitions of electrons from the valence band of the p—type semiconductor to the conduction band of the n—type semiconductor prevail (in Fig. 83, transitions from right to left). These transitions are unlimited, their number increasing with the reverse bias. This explains the rapid growth in the reverse current through the diode (region containing point 7 on the current-voltage characteristic in Fig,. 81).

5.3.6. Generation of Continuous Oscillations with the Help of the Tunnel Diode. We shall illustrate the application of the tunnel diodes using the generation of continuous oscillations by way of an example. The negative resistance of the tunnel diode can be used to compensate the positive resistance of a certain part of the electric circuit and to amplify

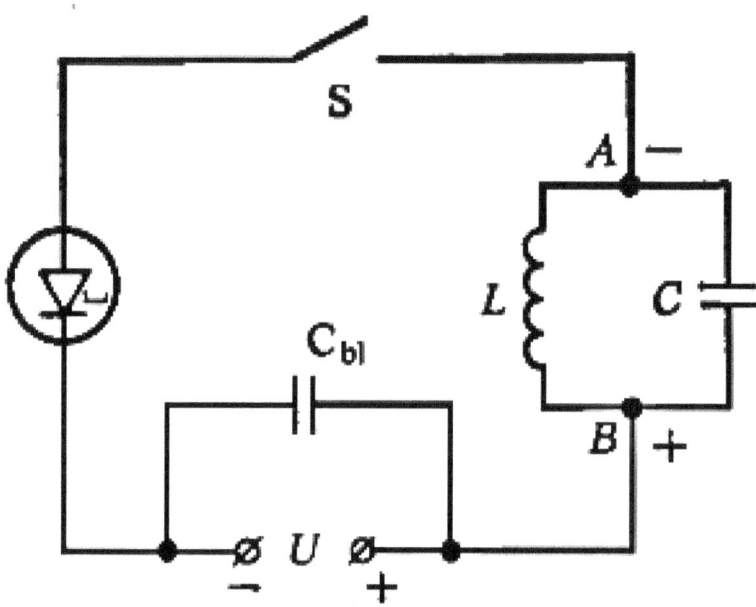

FIGURE 13

a signal or generate oscillations. For example, if the working point of a diode connected in series with an oscillatory d.c. circuit (Fig. 84) is on the descending region of the current-voltage characteristic, the energy losses in the oscillatory circuit are made up for, and continuous oscillations appear.

When the switch S is closed, small amplitude free oscillations appear in the oscillatory circuit, which in the absence of the tunnel diode would damp. Let us choose the supply voltage U such that the working point of the diode is at tho middle of the negative resistance region on the current-voltage characteristic. As electric oscillations are generated in the oscillatory circuit, polarities of points A and R alternate each half-period. During one half-period, the polarity of these points will be as shown in the figure. In this case, the voltage across the oscillatory circuit is subtracted from the supply voltage, and the total forward bias across the diode decreases. Since the diode operates in the negative resistance region in the regime chosen, a decrease in the forward bias will cause an increase of current through the diode, and hence in the entire circuit. On the other hand, when the polarity of the oscillatory circuit terminals changes (during the second half-period), the forward bias will increase, while the current in the circuit decreases. Hence, the current in the circuit pulsates. It can be easily seen that the alternating component of this current coincides in phase with the voltage oscillations in the circuit. This means that the power in the part of the circuit formed by the oscillatory circuit is positive ($\cos \varphi = 1$), and the energy is continuously supplied to the oscillatory circuit. Hence, the amplitude of oscillations in the circuit will increase. However, at the same time the energy losses also increase. When the equilibrium between the energy losses and replenishment is attained, continuous oscillations occur in the oscillatory circuit.

It can be seen from the diagram that the continuous-wave oscillator based on the tunnel diode is much simpler than a valve oscillator.

At present, tunnel diodes are widely used in , electronic computers as well as other radio-electronics circuits where high speed is required. This is because of their exceptionally small time lag (a tunnel transition of electrons through the potential barrier takes only from $10^{-12} to 10^{-14} sec$). This property of tunnel diodes makes them applicable for generation and amplification of microwave oscillations (with frequencies up to hundreds of gigacycles).

Tunnel diodes are also used as high-speed switches (the switching time can be reduced to sec). In an electric circuit, the tunnel diode operates as a rectifier which switches to the ON state as the forward bias decreases and to the OFF state as this bias increases.

5.3.7. Backward Diode. This is an interesting modification of tunnel diodes. The doping level of semiconductors used to manufacture them is somewhat lower than in the case of ordinary tunnel diodes (the concentration of impurity introduced into the semiconductor in this case is about $10^{18} cm^{-3}$). The Fermi levels in such semiconductors coincide with the boundaries of the allowed bands: the Fermi level in the $n-$type semiconductor coincides with the bottom of the conduction band and with the top of the valence band in the $p-$type semiconductor. If we consider the junction between two such semiconductors in equilibrium, it turns out that their energy bands do not overlap (Fig. 85). Hence, in the absence of an external bias tunnel transitions through the interface do not occur. The application of the forward bias does not induce these transitions either, since in this case also the allowed energies of electrons in one region are opposed by the forbidden band in the other region. For this reason, the forward current in the diode can appear only due to above-the-barrier transitions of carriers.

Since the potential barrier at the interface between semiconductors with such a high doping level is sufficiently large (it can be seen from the figure that it is equal to the forbidden band width), the forward current is found to be negligible up to quite high values of the forward bias (Fig.

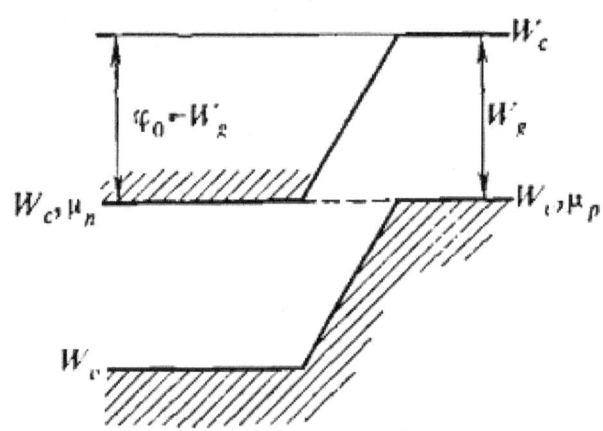

FIGURE 14

86). Practically, it is equal to the diffusion current typical of tunnel diodes in general (dashed line in Fig.81).

On the other hand, when the external voltage across the diode is applied in the reverse direction, the allowed bands overlap more and more as the bias increases. In this case, the tunnel transitions become possible, their number increasing unlimitedly with U_r, just as in the case of ordinary diodes. This leads to the rapid increase in the reverse current, which becomes much higher than the current in the forward direction. Thus, properties of this type of diodes with respect to the dependence

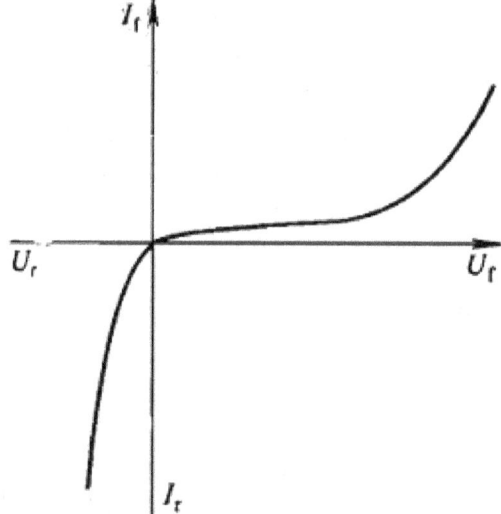

FIGURE 15

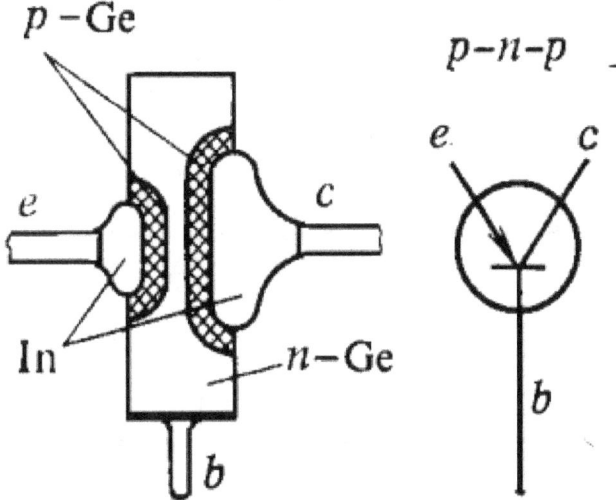

FIGURE 16

of conductivity on the bias voltage are opposite to properties of ordinary diodes, which explains their name backward diodes. Since their current-voltage characteristics do not have a negative resistance region, they cannot be used for generating or amplifying oscillations, but they can be used as detectors in the range of very high frequencies.

5.4. Transistors

In contrast to semiconductor diodes, transistors are semiconductor systems containing three regions separated by two $p-n$ junctions. Each of the regions has its own terminal. For this reason. in analogy with vacuum triodes, transistors are often called semiconductor triodes. Separate transistor devices are similar to vacuum triodes in their duty and are mainly used for the voltage and power amplification of electric signals.

In order to manufacture a transistor, an impurity with opposite type is introduced by alloying or diffusion into two opposite faces of a semiconductor single-crystal plate. Either the $p-n-p$ type or $n-p-n$ type transistor can be treated in this way. There is in principle no difference between them. As a matter of fact, holes play the major role in $p-n-p$ type transistors while electrons, in $n-p-n$ type transistors. Let us consider the $p-n-p$ type transistor.

5.4.1. $p-n-p$ Transistor. Figure 87 shows a schematic diagram of the transistor based on the n-type germanium wafer with two p-type

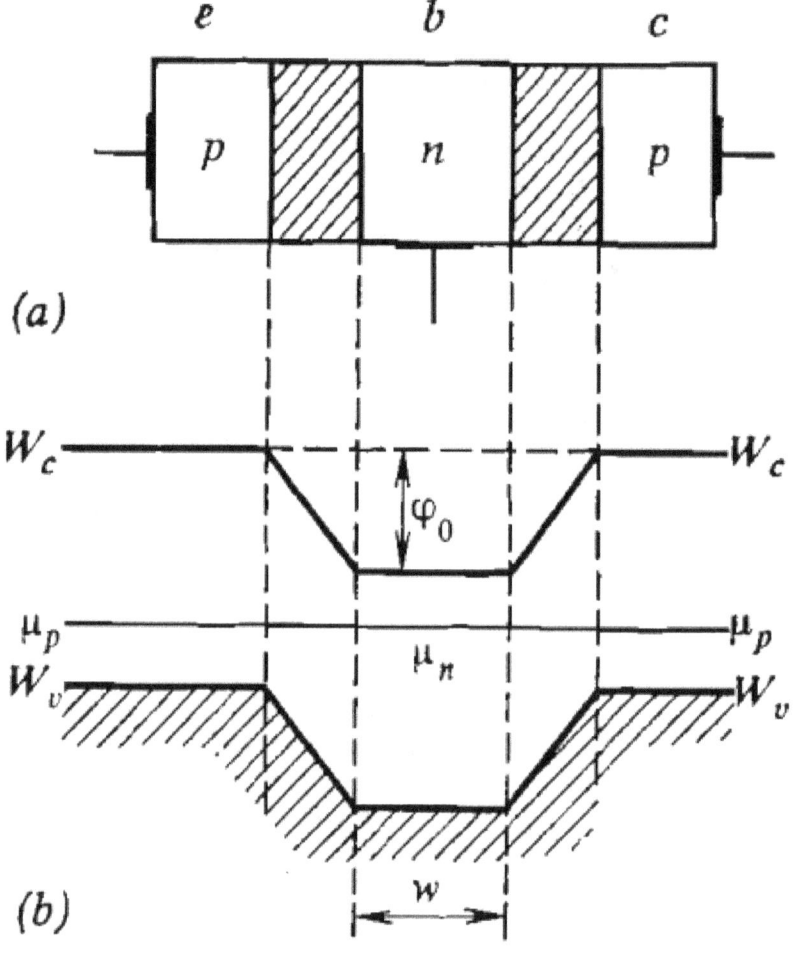

FIGURE 17

regions treated by indium pellets alloyed at certain places on both sides. The middle part of the transistor is called the base, and the two other parts are called the emitter and the collector. These parts are separated from the base by the emitter and collector $p-n$ junctions. The area of the collector junction is made such that it is larger than the area of the emitter junction. This ensures that carriers coming to the base from the emitter are efficiently trapped by the field of the collector junction. The junction transistor is shown schematically in Fig. 88.a. In equilibrium, when the power supply is not connected to the transistor, the resultant currents through the two junctions are equal to zero, and the Fermi levels of the three regions are at the same height.

In working conditions, the forward voltage is applied across the emitter $p-n$ junction ("+" is connected to the emitter and "−" to the base). This lowers the potential barrier and narrows the junction region. On the other hand, a reverse bias is applied to the collector $p-n$ junction (("+" to the base and "−" to the collector), and this increases the potential barrier and makes the junction wider (Figs. 89.a and b). Let us first consider processes occurring in each of the junctions separately.

5.4.2. Injection of Holes to the Base. The lowering of the potential barrier of the emitter junction results in an intense injection of carriers and in the diffusion current. The electron component of this current, which is determined by the transition of electrons from the base to the emitter turns out to be unnecessary for (and even harmful to) the transistor's operation. Hence, in practice it must be reduced to zero. This can be done if the doping level of the emitter is much higher than that of the base. Since the hole concentration in the emitter is much higher than the electron concentration in the base ($p_{po} \ll n_{no}$), the number of holes going from the emitter to the base greatly exceeds the number of electrons coming to the emitter from the base, and we can assume that the total diffusion current through the emitter junction is created only by holes injected from the emitter to the base.

The intense injection leads to a sharp increase in the concentration of holes (the minority carriers) in the base at the interface with the emitter.

Let us now turn to the collector junction.

5.4.3. Collector Junction. The forward bias is applied to the collector $p-n$ junction. Therefore (if the power supply is disconnected from the emitter junction), only the reverse current flows through it, and this will be equal to the saturation current i_s, and will be created solely by the minority carriers, viz. by holes moving from the n–type base to the collector and by electrons moving from the p–type collector to the base. For the collector junction, the electron component of the current is also useless. In this case, too, it can be suppressed by creating a sharp difference in the doping levels of regions in contact so that $p_{po} \gg n_{no}$. Under this condition, the number of electrons (the minority carriers) in the collector is relatively small, and we can assume that all the reverse current through the collector junction is created by holes moving from the base to the collector. Since the field of the collector junction entrains holes coming from the base and carries them to the collector, the hole concentration in the base at the interface with the collector is practically zero.

Let us analyze the situation when the two $p-n$ junctions operate simultaneously. Because the injection of holes from the emitter to the base is intense, the hole concentration in the base at the interface with

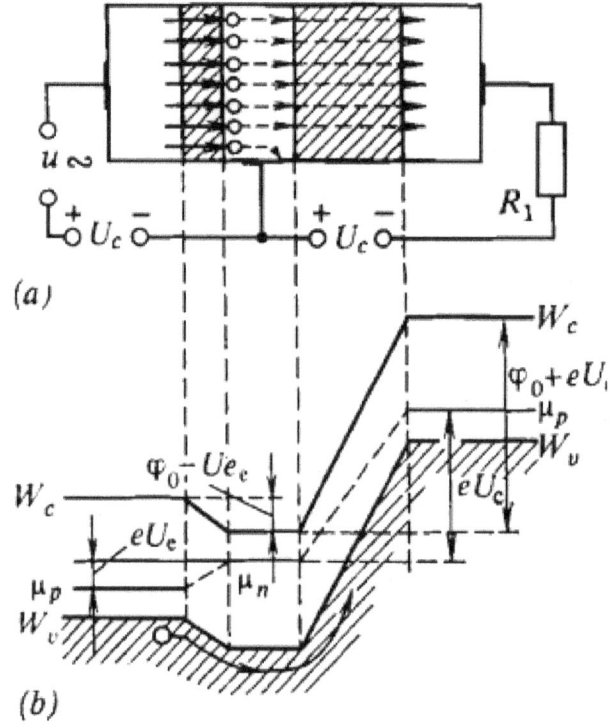

FIGURE 18

the emitter is many times higher than the equilibrium value, while at the interface between the base and the collector the hole concentration is close to zero. This sharp difference in concentrations leads to the intense diffusion of holes through the base from the emitter to the collector. The rate of this process is dependent on the small thickness of the base, since some holes traveling through the base recombine with electrons. The thinner the base, the fewer holes recombine with electrons in the bulk and the more holes reach the collector junction. If the base thickness w is much less than the diffusion length for holes in the n–type region ($w \ll L_p$), almost all the holes injected from the emitter to the base will reach the collector junction. Here, they are entrained by the junction field and are carried to the collector. Clearly, the additional collector current created by the transition of these holes is practically equal to the emitter current. The reverse current through the collector junction is much smaller than the current created by holes injected from the emitter to the base and continuing their motion through the collector junction. Therefore, the collector current can be taken as being equal to the emitter current ($I_c \simeq I_e$).

5.4.4. How Does the Transistor Amplify? The amplifying action of a transistor is based on the equality between the collector and the emitter currents. This action can be realized due to the large difference in the resistances of the collector and the emitter $p-n$ junctions when connected in opposite directions.

The forward biased emitter junction has a low resistance, and the voltage drop U_e across it is small. On the other hand, the collector junction is reverse biased and has a much higher resistance. Consequently, a high resistance load can be connected to the collector circuit. This resistance R_1 is much higher than the resistance of the emitter junction. Since the emitter and collector currents are equal, the voltage drop $U_1 = I_c R_1 \simeq I_e R_1$ across the high-resistance collector load is much larger than the voltage drop U_e across the emitter junction ($U_1 \gg U_e$). This is the essence of the voltage amplification effect of the transistor.

Since $U_1 I_c \gg U_1 I_e$ power amplification also v. takes place: the output power P_{out} on the high-resistance load in the collector circuit is much higher than the input power P_i supplied to the emitter junction ($P_{out} \gg P_{in}$)

5.4.5. Methods of Connecting a Transistor. There are three different ways of connecting a transistor in a circuit: common base connection, common emitter connection, and common collector connection.

When the common base connection is used (Fig. 90.a), both the emitter and the collector voltages are measured with respect to the base. The emitter circuit is then the input circuit, while the collector circuit is the output circuit.

The amplifying action of a transistor with this type of connection (see also Fig. 91.a) has been discussed above.

Figure 90.b shows the common emitter connection of a transistor. In this case, potentials of the base and the collector are measured with respect to the emitter. The base circuit is the input, while the collector circuit is the output one. This type of connection is the one most widely used in transistor devices.

In the case of the common collector connection (Fig. 90.c) the input signal is applied to the collector-base junction, and the load resistor is connected between the emitter and the collector. This type of connection is not used as often as the previous two types.

5.4.6. Common Emitter Connection. Let us consider an amplifier stage based on the $p-n-p$ transistor connected using the common emitter method (Fig. 91.b). The source of the signal being amplified is connected to the emitter-base junction circuit in series with the bias voltage source.

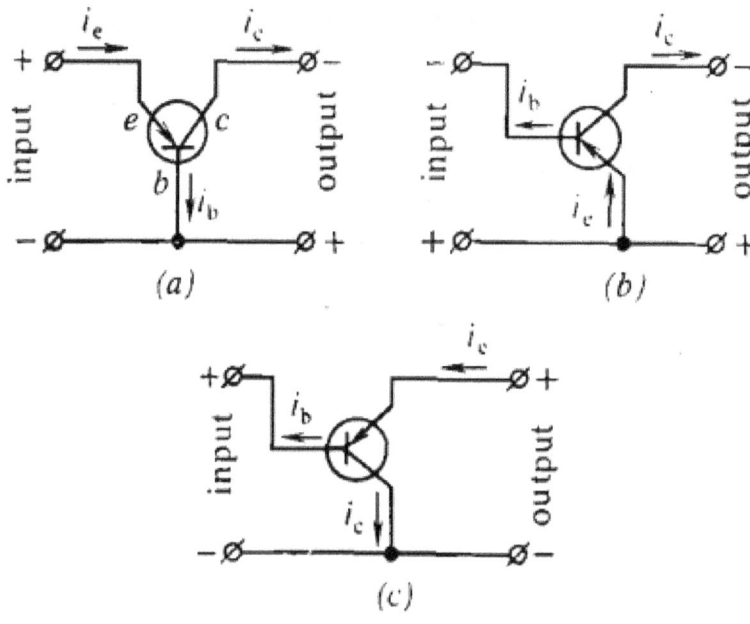

FIGURE 19

Since tho emitter junction is forward biased, tho magnitude of the input resistance is comparatively small.

The power supply is connected in the emitter-collector circuit in such a way that for the collector junction its voltage is reverse. With such a connection, the resistance of the collector junction is large making it possible to connect a high-resistance load R_l in the collector circuit, from which the amplified signal can be taken.

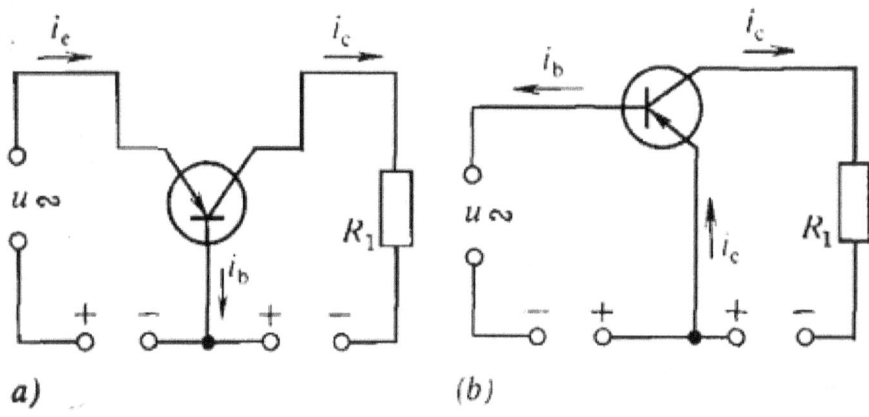

FIGURE 20

The voltage of the signal being amplified causes a change in the current of the emitter-base circuit, which in turn alters the resistance of the collector junction and hence leads to a redistribution of the voltage drop of the source between the transistor and the load. Indeed, during the half-period when the voltage of the signal being amplified is added to the forward bias voltage across the emitter junction, the injection of carriers from the emitter to the base is intensified, resulting in an increase in the emitter as well as the collector currents. On the other hand, the arrival of movable carriers (holes) at the depletion region of the collector junction decreases its resistance, due to which the greater part of the supply voltage applied to the collector circuit. drops on the load resistance R_l. A slight increase in the forward bias across the emitter junction is accompanied by a considerable increase in the potential difference across the load resistance R_l. When the input signal is of the reverse polarity, the opposite situation takes place: the resistance of the collector junction increases and the voltage across the load decreases.

If the input signal is sinusoidal in form, the amplitude of the alternating voltage component across the load resistor R_l is found to be dozens and even hundreds of times higher than the amplitude of the alternating input voltage. Unlike a common base connection, a common emitter operation yields considerable current amplification. This is because the change in the collector current (practically equal to the change in the emitter current) is many times greater than the change in the base current (the input current of the amplifier stage). Current (like voltage) is amplified dozens and hundreds of times. Simultaneous current and voltage amplifications provide the best opportunity of power amplification. The power amplification factor, defined as the ratio of the output power to the input power may reach several thousand and even tens of thousands in the case of common emitter connection.

5.5. Semiconductor Injection (Diode) Lasers

5.5.1. Photons Create Photons. It was shown above that semiconductors can absorb as well as emit light.

The absorption of light with the energy of quanta $h_v \simeq W_g$ is primarily associated with the photo-conductive effect. By absorbing photons, electrons from the valence band move to the conduction band (Fig. 92.a), i.e. go from an unexcited state to an excited state. They do not remain in this state for long. After a certain time (which depends on the forbidden band width of the semiconductor), they return to the valence band, recombining with holes and emitting light quanta with an energy

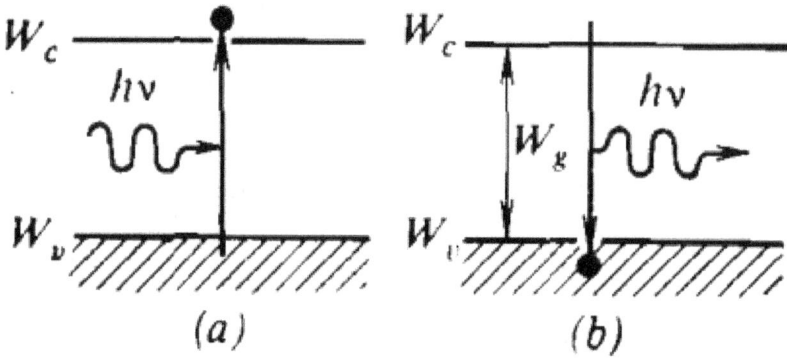

FIGURE 21

$h\nu$ approximately equal to the forbidden bandwidth W_g (Fig. 92.b). Unless the superconductor experiences an external influence, electrons return to the valence band impulsively, or spontaneously. Nobody can predict when spontaneous recombination will occur and which properties the emitted photon will have.

However, in addition to spontaneous recombination, induced recombination is also possible due to the effect of external radiation. A photon getting into a semiconductor and meeting an excited electron on its way may "push" it and thus force, it to return to the valence band. In this case, the primary photon acting on the electron does not change its properties, while the photon emitted during the recombination between the electron and a hole is identical in its properties to the primary photon: it has the same frequency and energy, the same direction of propagation and polarization. At the exit from the semiconductor, the primary photons cannot be distinguished from photons of the induced radiation (Fig. 93.a, b).

5.5.2. Population Inversion. A photon getting into a semiconductor can either be absorbed and vanishes, thus creating free electron and a hole, or cause a recombination accompanied by the generation of a new photon. The probabilities of these two processes are equal. This does not mean, however, that the two processes occur with equal intensity. The ratio between the number of electron transitions from the valence band to the conduction band (absorption of light) and the number of reverse transitions (emission of light) depends not only on the probability of the individual transitions but also on the total number of electrons which can take part in each of these transitions. In other words, the number of transitions accompanied by absorption of light is proportional to the number of electrons populating the upper energy levels near the top of the valence band, while the number of transitions associated with the

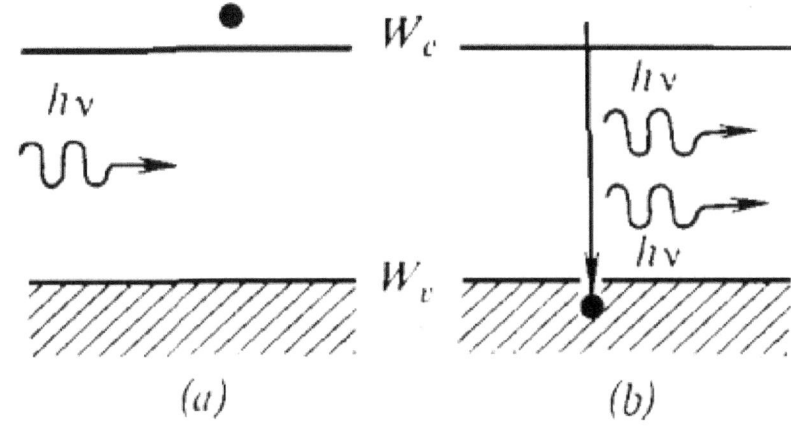

FIGURE 22

emission of light is proportional to the number of electrons occupying the lower energy levels near the bottom of the conduction band. In the general case, when the distribution of electrons among the energy states is determined only by thermal excitation, the population density of the upper levels of the valence band is always much higher than the population density of the lower levels of the conduction band. Therefore, in normal conditions a semiconductor absorbs light.

In order for a semiconductor to amplify, rather than absorb light, it is necessary that the electron population density of levels adjoining the bottom of the conduction band be higher than the population density of levels near the top of the valence band. In this case, with equal probabilities of the creation and recombination of electron-hole pairs, the total number of recombination will prevail. This distribution of electrons among the energy levels is the inverse to their distribution at thermal equilibrium. For this reason, a semiconductor in this state is said to be characterized by population inversion.

A semiconductor in which most of the levels near the bottom of the conduction band are occupied by electrons is a degenerate n—type semiconductor with the Fermi level in the conduction band proper. On the other hand, a semiconductor with a low electron population density of levels adjoining the top of the valence band (in other words, with a high hole population density of these levels) is a degenerate p—type semiconductor with the Fermi level lying in the valence band. Thus, a transition of a semiconductor to the active state characterized by the population inversion is associated with the creation of the simultaneous degeneracy in both electrons and holes (Fig. 94). In this case, the difference be-

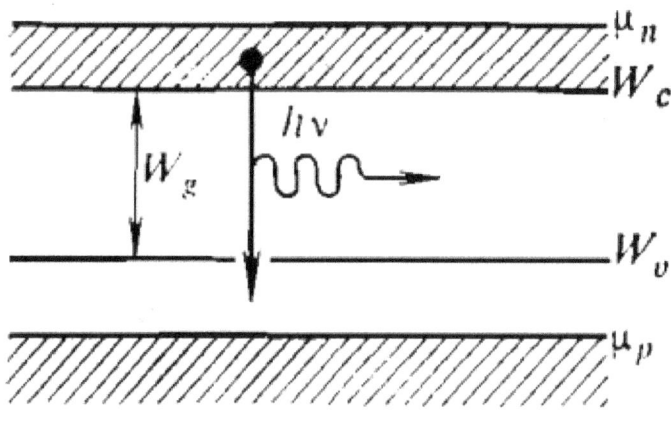

FIGURE 23

tween the Fermi levels μ_n for electrons and μ_p for holes turns out to be larger than the forbidden band width:

$$\mu_n - \mu_p > W_g$$

This relation is called the population inversion condition for a semiconductor.

The semiconductor with the simultaneous degeneracy in electrons and holes is ideal for amplification of light whose energy of quanta ranges between $h\nu_{min} = W_g$ and $h\nu_{max} = \mu_n - \mu_p$. Indeed, light quanta whose energy belongs to this range will not be absorbed by such a semiconductor because on the one hand, there is nothing to absorb them (the number of electrons at the top of the valence band is small) and, on the other hand, even the small number of electrons which may appear in this region have no vacancies to go to upon the absorption of a quantum (practically all the levels at the bottom of the conduction band are filled by electrons). At the same time, this state creates favorable conditions for recombination of electrons occupying the filled levels of the conduction band with holes in the empty levels of the valence band. During this recombination, the quanta of light with the energy in the indicated range are emitted.

5.5.3. Creation of Population Inversion at the Junction Between Degenerate Semiconductors. There are different ways of creating the population inversion in a semiconductor: irradiation of a semiconductor by light, bombardment by fast electrons, or direct electric excitation. We shall discuss the most interesting method, viz. the creation of the population inversion at the $p-n$ junction between two semiconductors

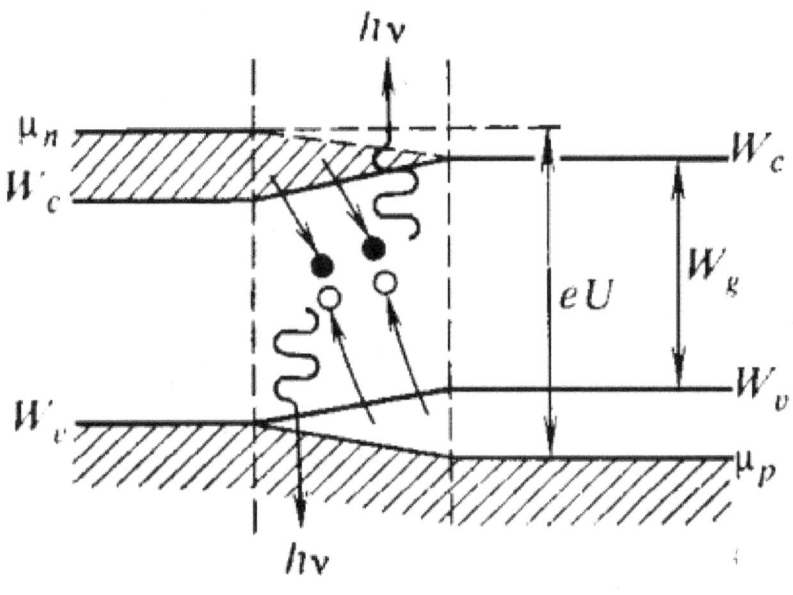

FIGURE 24

one of which is degenerate in electrons and the other in holes. Since the difference between the Fermi levels of such semiconductors is greater than W_g (see Fig. 80), the potential barrier $\varphi_0 = \mu_n - \mu_p$, appearing during the formation of the $p-n$ junction is found to be larger than the forbidden band width W_g. If we apply a forward voltage to the junction comparable with the contact potential difference, the potential barrier will vanish, and a region will appear near the interface where the population inversion condition is observed (Fig. 95):

$$eU = \mu_n - \mu_p > W_g.$$

A sharp lowering of the potential barrier leads to an intense injection of electrons from the n–type region and holes from the p–type region to the $p-n$ junction. When the opposite fluxes of these carriers meet at the $p-n$ junction, they recombine and emit light. The higher the applied potential difference, the larger the current flowing through the junction and the more intense the recombination. The minimal current at which the intensity of the recombination radiation becomes comparable with the intensity of light absorbed by the $p-n$ junction is called the threshold current. At a current exceeding the threshold value, the $p-n$ junction becomes an active medium which amplifies the light propagating in the plane of the junction. The $p-n$ junction in this state can be used as a quantum-mechanical amplifier of light, the energy of whose quanta is close to the forbidden band width. The primary radiation being amplified need not impinge on the $p-n$ junction from outside; it can be created in the junction itself by spontaneous transitions. In this case, the $p-n$ junction operates as a quantum-mechanical oscillator, or laser. Since the operation of such a source of light is based on the injection of carriers into the $p-n$ junction, this type of oscillator is called the semiconductor injection laser.

In order to improve the oscillation conditions, all laser systems have feedback mechanisms to return some of emitted radiation back to the active medium. Usually, semitransparent mirrors are used for this purpose. In semiconductor lasers the parallel polished faces of the crystal itself, which are perpendicular to the plane of the $p-n$ junction, act as mirrors. Some of the light quanta reflected by the output face and returned to the crystal cause induced recombination as they pass through the crystal, and this is accompanied by the emission of additional light quanta identical to the reflected ones. A further reflection by the other face results in a still greater amplification of transmitted light, and so on. With multiple reflections, a very large amplification of the radiation propagating in the plane of the $p-n$ junction and perpendicular to the polished reflecting faces of the crystal can be attained (Fig. 96).

The most commonly used injection lasers are those based on gallium-arsenide (Ga As). The crystal faces in these lasers have sizes of the order of $0.2-1\,mm$, and the thickness of the $p-n$ junction is about $0.1\,\mu m$. The thickness of the emitting layer is somewhat larger than the junction thickness and reaches $1-2\,\mu m$. The output power of Ga As lasers is of the order of tens of watts. Injection lasers have high efficiency (defined as the proportion of the electric energy converted into light) and this quantity often exceeds $50-60$. Their high efficiency, small sizes, simple design and considerable output power make injection lasers very promising devices. The most important fields of application of injection lasers are computers, radar sets with a high resolving powers, optoelectronic systems, wireless transmission lines, and television. Logic elements (gates) based on injection lasers can be used to construct a purely optical computer working at 10^9 operations per second or higher.

5.6. Semiconductor at Present and in Future

semiconductors burst impetuously into 20th century science and technology. Their extremely low energy consumption. the remarkable compactness of devices due to very dense packing of elements in circuits, and high reliability mean they are the most important materials in electronics, radio-engineering and science. Space exploration would have been impossible without semiconductor devices, since the requirements of small size, low weight and energy consumption are especially stringent for the spacecraft equipment. By the way, in space the energy is supplied from solar batteries that use semiconductor elements.

Microelectronics has opened new prospects for the development of semiconductor technology. It has become possible to replace semiconductor devices assembled from separate elements by the integrated circuits. Modern technology allows designers to create 10^5 elements per cm^2 of surface. By using the layered structures developed during last years (metal-nitrite-dielectric-semiconductor), 10^8 circuits that can store information and can be placed per mm^2 of surface. The information can he read out not only with the help of electric signals but also by using laser radiation obtained, for example, from semiconductor injection lasers.

However, the potential of semiconductors has far from being exhausted and these await new researchers.

www.ingramcontent.com/pod-product-compliance
Lightning Source LLC
Chambersburg PA
CBHW080914170526
45158CB00008B/2112